fourth symposium on

the structure of low-medium mass nuclei

1. B. Wildenthal
2. T. T. S. Kuo
3. E. Halbert
4. R. E. Segel
5. J. A. Becker
6. S. J. Krieger
7. P. M. Endt
8. S. S. Hanna
9. J. P. Davidson
10. N. Koller
11. E. Titterton
12. R. W. Krone
13. M. Harvey
14. H. Hennecke
15. S. Maripuu
16. B. Castel
17. R. Horoshko
18. R. T. Carpenter
19. F. E. Dunnam
20. R. G. Arns
21. A. K. Hyder
22. G. Engelbertink
23. G. G. Seaman
24. M. L. Roush
25. G. B. Beard
26. C. M. Class
27. F. E. Durham
28. Chia-Cheh Chang
29. D. R. Tilley
30. F. W. Prosser, Jr.
31. A. Rollefson
32. J. Umbarger
33. H. T. Fortune
34. D. Kurath
35. P. Brussaard
36. H. Bakhru
37. B. D. Kern
38. L. Seagondollar
39. S. A. Moszkowski
40. R. Santo
41. J. Dubois
42. H. M. Kuan
43. A. J. Howard
44. L. Zamick
45. R. Lawson
46. C. E. Moss
47. P. Goldhammer
48. R. W. Finlay
49. S. Fiarman
50. R. C. Bearse
51. J. R. McDonald
52. E. K. Warburton
53. G. I. Harris

fourth symposium on
the structure of low-medium mass nuclei

Edited by
J. P. Davidson

Sponsored by:
The Nuclear Structure Laboratory
The University of Kansas
with support by the U. S. Atomic
Energy Commission and the
University of Kansas

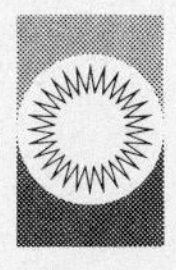

University Press of Kansas
Lawrence / Manhattan / Wichita

Standard Book Number 7006-0089-2

Library of Congress Catalog Card Number 71-190081

Printed in the United States of America

1057

EDITOR'S PREFACE

This second published volume of the proceedings, as the previous one, contains all of the invited papers, formal comments and discussion following these which were presented at the fourth Symposium held at the University of Kansas on October 12-14, 1970. Dr. Robert C. Bearse of the Nuclear Structure Laboratory was the co-organizer of the Symposium. These proceedings are in form and style much as before and they differ in but two important respects from the earlier proceedings. The invited papers are denoted by the Roman numeral of the session and a capital letter, while the formal comments are denoted by the session numeral and a lower case letter. The discussion was tape recorded and immediately transcribed. All of these transcriptions were examined and corrected by those commenting. Again editorial insertions, where they were thought useful or necessary, are enclosed in square brackets.

Of the differences between this and the previous volume the least important is that all of this volume was typed from manuscripts furnished by the authors; however, they did not see proof of the retyping so that typographical errors are ours. The most serious difference between this and earlier such proceedings is that one contribution here is not based upon a manuscript furnished by its author but on a transcription of the original tape recorded talk. The speakers have always been invited with the understanding that they would provide a proper manuscript upon completion of their talk or very soon thereafter. In the past the threat of using the recorded talk has been sufficient encouragement so that the speakers have always departed the Symposia leaving their manuscripts in the hands of the editor. Unfortunately this time Professor Endt, making his fourth appearance, failed to provide such a manuscript or indeed respond to various communications to him. As editor I had to choose between a complete record of the Symposium with an imperfect paper of one of the talks, or an imperfect record. I decided upon the former. Since Professor Endt had a copy of the typescript of his recorded talk for more than a month, one must conclude that he considers it a faithful rendition of his remarks--in any event the scientific responsibility for it is his and his alone. It is a pity for as the discussion shows it was an interesting, stimulating and well received talk. Furthermore, while these papers vary greatly in length (8 pages to 46 pages) they all represent at least one hour talks.

Finally, we again devoted a session to industrial speakers and the subject was the use of computers in scientific data acquisition. All of the major computer manufacturers were invited to make presentations and we were perhaps fortunate that only two were able to attend, since we could only devote three hours to such a session.

For concerned examiners of frontispieces, I am happy to report that Dr. Harvey is much improved and no doubt will return to the Deep River ski slopes come winter.

Our sincere appreciation goes to several of the graduate students of the Nuclear Structure Laboratory, Messrs. Boudrie, Close, Kanatas, Lockwood, Maharry, Shirk, Sievers and Wilson who tape recorded the talks and discussion and helped translate the latter to our able typist Nancy Grant. I wish to thank not only Miss Grant but Mrs. Susan Bryan who ably and indeed cheerfully typed and retyped the entire manuscript until it was properly done.

I also wish to acknowledge the role played by Francis H. Heller, Vice Chancellor for Academic Affairs whose understanding of the importance of these Symposia to the Nuclear Physics Community has been instrumental in providing the generous support we have received from the University of Kansas. Finally, I want to express my appreciation to John H. Langley, Director of the University Press of Kansas, and his able staff whose efforts have made this project again a success.

Lawrence, Kansas
July, 1971

J. P. Davidson

CONTENTS

I.A. THE COLLECTIVE PROBLEM IN THE 2s-1d SHELL

M. Harvey
Atomic Energy of Canada, Limited
Chalk River Nuclear Laboratories
Chalk River, Ontario, Canada

The nuclear 2s-1d shell is particularly rich in exhibiting various kinds of collective phenomena including rotational bands implying stable, intrinsic states, quadrupole and octupole vibrations and of course the giant dipole resonance indicative of the collective movement of neutrons against protons. In this talk I wish to discuss some of the studies that I have been associated with in firstly recognizing low energy collective phenomena using qualitative macroscopic models and, secondly, trying to understand how these collective features come about in the microscopic models. Our philosophy has been to look at data from complex nuclei and ask what it can tell us about the general structure of the nucleus and the form of the effective interaction. As you will see our approach is very analytical.

About 10-15 years ago the discovery of rotation-like bands in such nuclei as Mg^{25}, F^{19}, Mg^{24} and Ne^{20} excited considerable interest[1]. The fascination was in the fact that nuclei with so few particles should be able to exhibit collective phenomena. The discovery gave one hope that one could understand collective features in a microscopic way from the independent particle model which, hitherto, had been the sole means of studying light nuclei. Probably the most elegant bridge between the collective and independent particle models was built by Elliott[2,3]. He showed how shell model states could develop rotational bands by classifying them simultaneously according to the LS-coupling and the symmetry group SU_3 of the harmonic oscillator. The effective interaction which is diagonalized in this coupling scheme is the now familiar QQ potential which generates bands of levels within which the energy is proportional to L(L+1) where L is the *orbital* angular momentum. The QQ-potential can actually be derived from the simple physical argument that it is the long range part of the effective interaction that will generate collective features.

There are subtle differences between the SU_3-rotation-like bands and those of the semi-classical rotational model[4]. In the first place the SU_3 scheme is a classification of configurations in the basic shell model and hence the bands can have only a finite number of members i.e., the bands appear "cut-off". Secondly the B(E2) transitions between members of an SU_3-band are not only proportional to the square of a Clebsch-Gordan coefficient, but also depend on other angular momentum terms[5], these latter terms tend to decrease the B(E2)'s between states near the band cut-off[6]. There is considerable interest in discovering whether the bands in light nuclei follow the characteristics of the semi-classical rotator or the quantum-mechanical, shell structured SU_3-model.

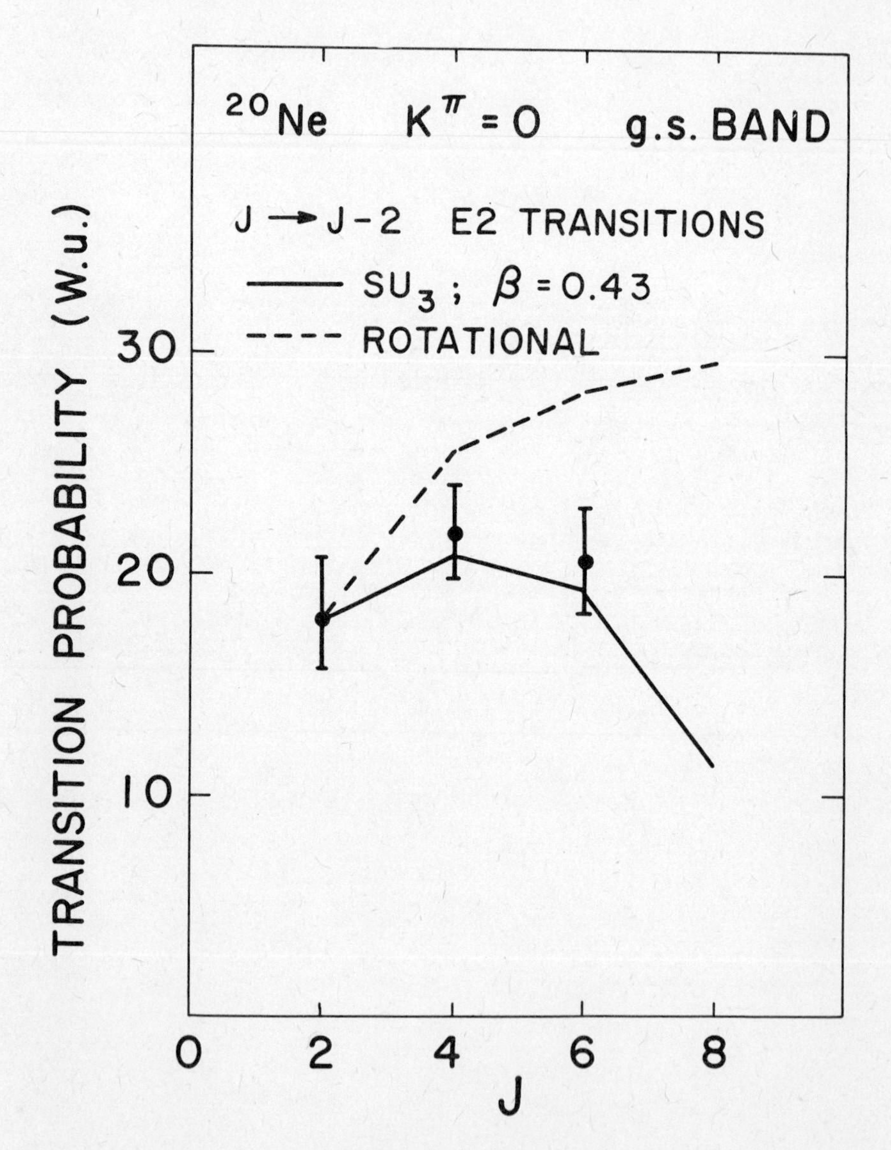

Fig. 1. B(E2:J→J-2) in the ground state band for Ne^{20} experimental points taken from Ref. 8.

Experiments have been performed to try to find states beyond the band cut-off for example[7]. Even this is not enough however since it is necessary also to show that the states beyond the band cut-off are in some way connected to the rest of the band - by an enhanced E2-transition for example. Experiments are also being performed to see whether the systematics of B(E2) transitions within a band follow the trend of the classical-rotator or the SU_3-model. Old data on Ne^{20} seemed to indicate that B(E2)'s were going to follow the classical model; however, new information from a Chalk River/Toronto collaboration[8], Figure 1 suggests that B(E2: $4 \rightarrow 2$) = 21.9 ± 2.1 W.U. (i.e. larger than before) and B(E2: $6 \rightarrow 4$) = 20.6 ± 2.4 W.U. (i.e. smaller) which is closer to the SU_3-predictions. It seems that the rotational bands in the 2s-1d shell are shell-model bands then and a cut-off is to be expected.

Despite the elegance and popularity of the SU_3-model, it can only suggest systematic features on even-even nuclei that exhibit rotation-like bands. The restriction to even-even nuclei arises because bands in SU_3 are in the orbital space and hence can only be compared with experiment when the total intrinsic spin S=0. To say anything at all about other nuclei we have to consider parts of the residual interaction other than the Q.Q. The remainder of the effective interaction consists of the short range part of the central force (which we can simulate by the pairing force) the exchange, and the spin dependent forces: these latter consist of the single particle spin-orbit force and the two-body spin-vector and spin-tensor.

If the two-body effective interaction remains essentially unchanged throughout the shell, then the spectral calculation for the n-hole configurations is equivalent to that for the n-particle except possible for a change of the single particle force: the two-body interaction remains invariant under this particle-hole transformation. If the single-particle potential is relatively unimportant then one should see a symmetry between spectra at the beginning and the end of the shell. This we apparently do not see. Thus, it is of interest to study the effect of the single particle potential on a spectrum which is dominated by the Q.Q-potential.

In Figure 2 we compare the single particle spectrum for O^{17} and the single hole spectrum for Ca^{39}. It is clear that for both particles and holes the s-state is below the d-state by about 1 MeV. Also the spin-orbit splitting (from $-x\ \vec{s}.\vec{\ell}$) is about the same except for a change of sign: in O^{17} (x = 2 MeV) the $d_{5/2}$ particle state is about 5 MeV below the $d_{3/2}$ state whereas in Ca^{39} (x = -2.5 MeV) the $d_{5/2}$ hole state is about 6.5 MeV above the $d_{3/2}$ state. Thus we wish to pay special attention to the effect of the single particle spin orbit force.

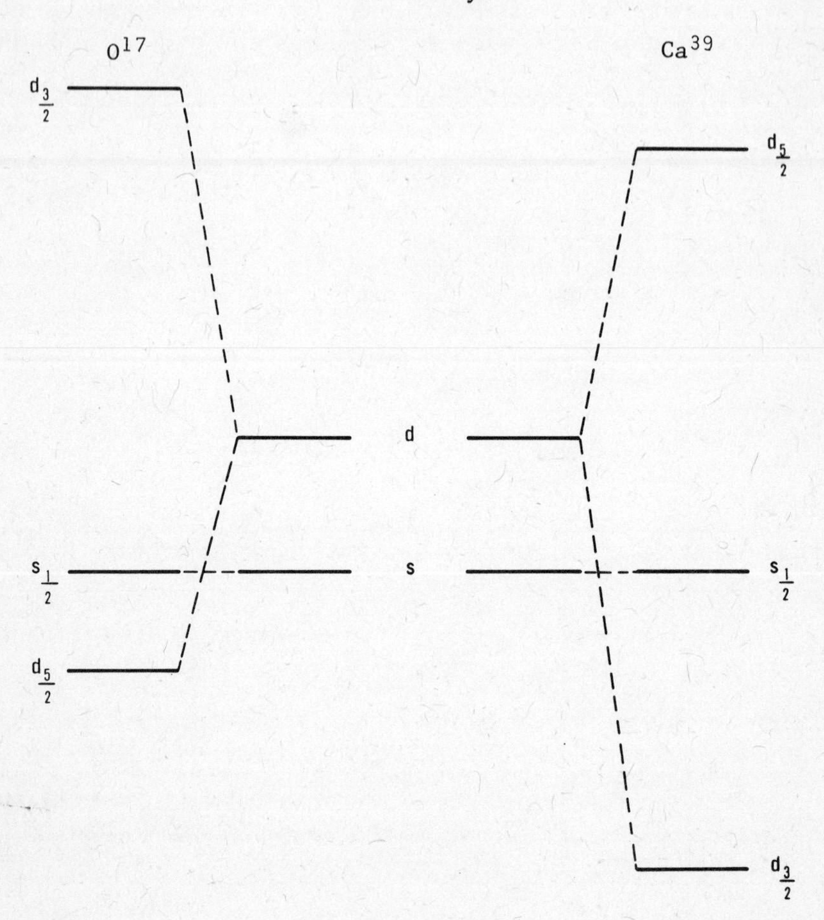

Fig. 2. Single particle spectrum for O^{17} compared with the single-hole spectrum for Ca^{39}.

Although there is an asymmetry between the properties of nuclei at the beginning and the end of the shell, the nuclei at the end still apparently exhibit collective features - this time of a spherical vibrational nature. In Figure 3 we show the recent data from a Chalk River/Queens University collaboration[9] in which the B(E2)'s between the lowest states in S^{32} were measured. The results bear out the expectation from the energy spectrum that S^{32} is a near harmonic-vibrating spherical nucleus. With the exception of the $B(E2:2^* \rightarrow 2)$ the B(E2) decays of the "second-phonon" states to the "one-phonon" 2^+ state are approximately twice that of the B(E2) from the one-phonon state to ground. That S^{32} is not a perfect

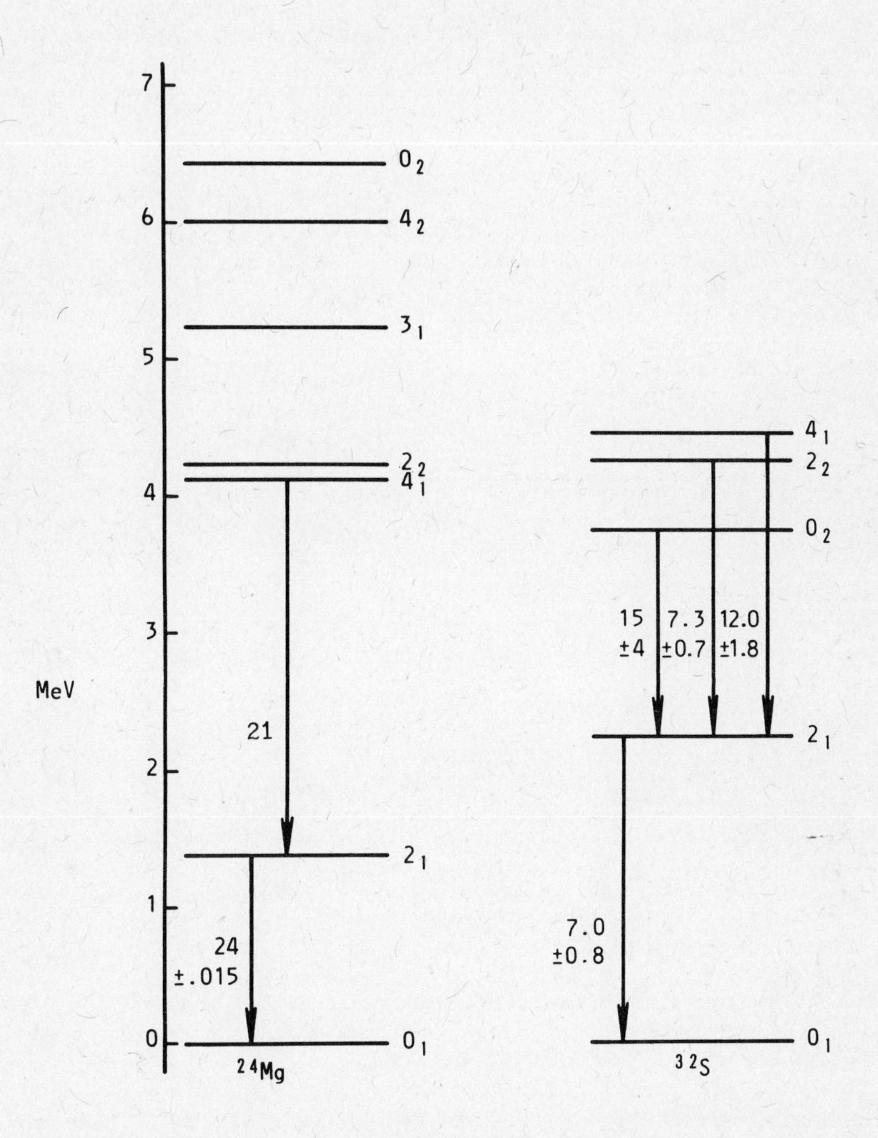

Fig. 3. B(E2)'s between low energy states in Mg^{24} and S^{32}. The S^{32} data is taken from Ref. 9 and that of Mg^{24} from Ref. 19.

vibrator is emphasized by the non-degeneracy of the second phonon 0^*, 2^* and 4^* triplet, the small $B(E2{:}2^* \rightarrow 2)$ and the non-zero quadrupole moment of the first phonon state. This latter has the value $Q(2^+) = -0.2 \pm 0.06$ barns from Nakai *et al.*, and a recent Chalk River experiment tends to confirm this.

This quadrupole moment is very large and, in the deformed rotor model would imply a prolate shape i.e. in contradiction to all Hartree-Fock calculations in the shell.

The assumed near harmonicity of S^{32} also seems to explain some of the gross features in the neighbouring odd-A nuclei.

Recently Castel and Stewart at Queen's University[12] have been extending the weak coupling model[13] for S^{33} and C^{33} by assuming the coupling of quasi-particles to an anharmonic-vibrating spherical S^{32} core. The Hamiltonian of the system has three terms

$$H = H_c + H_{sp} + H_{int}$$

$$H_c = \hbar\omega \sum_{\mu} b_\mu b_\mu + (1/4) \sum_{J=0,2,4} \hbar\omega\, \eta_J (b_\mu^* b_\mu^*)^J (b_\mu b_\mu)$$

$$H_{sp} = {}_{1/2}\, a^+_{s_{1/2}} a_{s_{1/2}} + {}_{3/2}\, a^+_{d_{3/2}} a_{d_{3/2}}$$

$$H_{int} = -\xi\hbar\omega\, (\pi/5)^{\frac{1}{2}}\, (Q_c \cdot Y^2)$$

Here the Hamiltonian for the core (H_c) includes a harmonic term with frequency $\hbar\omega$ and an anharmonic term that splits the degeneracy of the second phonon states: the splitting (values for η_J) is taken from experiment. Matrix elements of the core quadrupole operator Q are deduced from the observed B(E2: 2 → 0) of the core the quadrupole moment of the first phonon is *not* assumed to be zero. The parameters are chosen to get a best fit with experiment. The quasi-particle amplitudes are taken from the spectroscopic factors. In Figure 4 we show the type of spectrum one gets from such a scheme and the effects of the anharmonicity and the quasi particle amplitudes. I think it is clear that the non-harmonic structure of the core and the quasi-particle nature have a non-trivial effect on the resulting spectrum and results in a slight improvement over the original weak coupling models. In Figure 5 we show specifically the effect of taking a non-zero value for the first phonon quadrupole moment Q. It is seen that a non-zero, negative Q is indeed needed to get the ordering of levels at 4 MeV correct and the size is about that measured for the quadrupole moment for S^{32}. To say anything more about the structure of the core however, we have to return to more quantitative models.

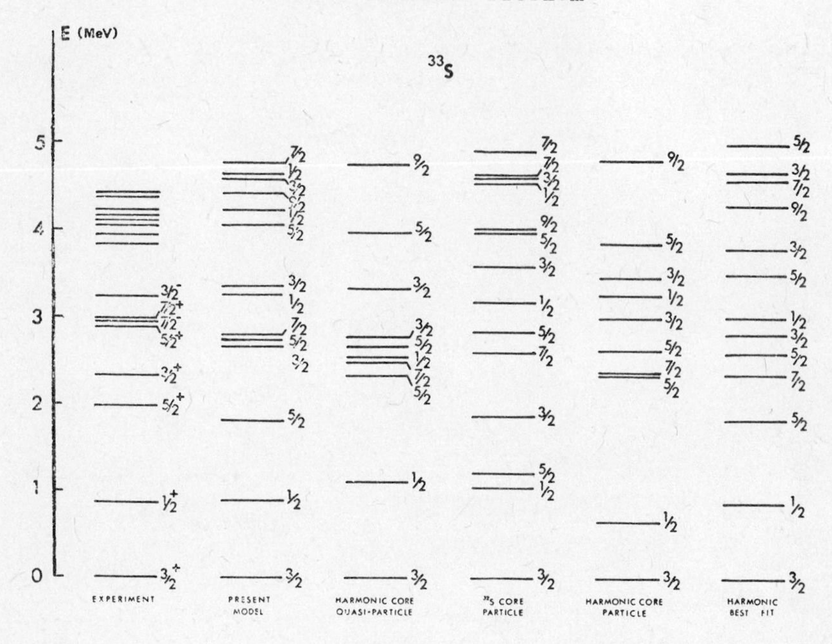

Fig. 4. The observed S^{33} spectrum in column 1 is compared in column 2 with a model calculation consisting of the weak coupling of quasi particles in $s_{1/2}$ and $d_{3/2}$ orbits to an anharmonic-vibrating core. Column 3 shows the effect of taking only a harmonic core with quasi particles; column 4 the anharmonic core with particles; and column 5 the harmonic core with particles. Column 6 shows the best fit obtained with the harmonic core with weakly coupled particle (*c.f.* Ref. 12).

Let me return now to the shell model and the investigation of the effect of the spin-orbit force. It should be apparent by now that our aim has been to try to correlate the onset of vibrations with the effect of the spin-orbit force. I tell you about the aim because it will explain why we have analyzed our models in the way we have. As you will see there is some evidence for the correlation but unfortunately I don't think it is striking enough for us to make any definite statement and clear up the problem once and for all.

In some recent calculations with S.S.M. Wong, from Toronto (see also Ref. 14), we set out to find the effect of the spin-dependent forces from effect of the spin-dependent forces from a realistic interaction in a pure shell model calculation for four fermions in the 2s-1d shell. For the calculation we took the Kuo[15] matrix

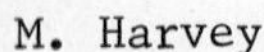

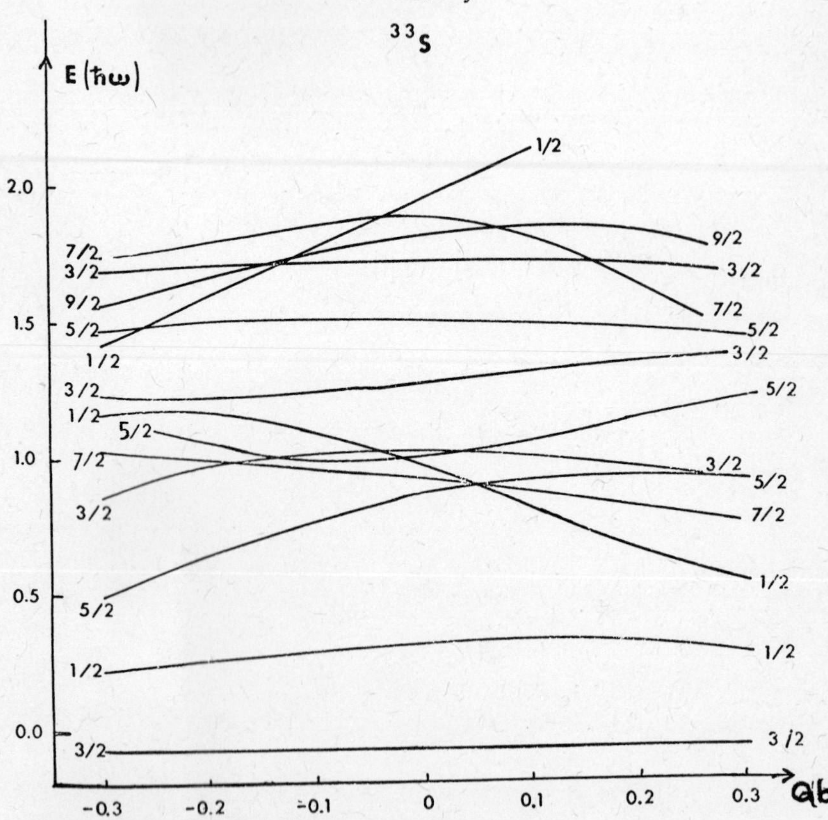

Fig. 5. The S^{33} spectrum in model of the weak coupling of quasi particles to an anharmonic core shown as a function of the quadrupole moment of the first excited state of the core. Best fit to experiment has Q = -.02 barns, i.e., an order of magnitude smaller than experiment in Refs. 10 and 11 (Ref. 12).

elements and, from these, computed the matrix elements for the constituent central, vector and tensor parts. We then ran the Rochester-Oak Ridge shell model program[16] with and without the various parts of the interaction to see what would happen. In Figure 6 we show the results of this calculation. For each group of four spectra the strength of the spin orbit force is varied among x = 2 MeV (appropriate for Ne^{20}), x = 0, x = -2, and x = y = -2.5 MeV (appropriate for Ar^{36}). For the first set on the left-hand side the two-body matrix elements consist only of those from the central force; in the next set we have added the two-body vector; and in the next we have the full potential which includes the tensor force. The low energy spectra of Ne^{20} and Ar^{36} is shown on the right. In looking for a

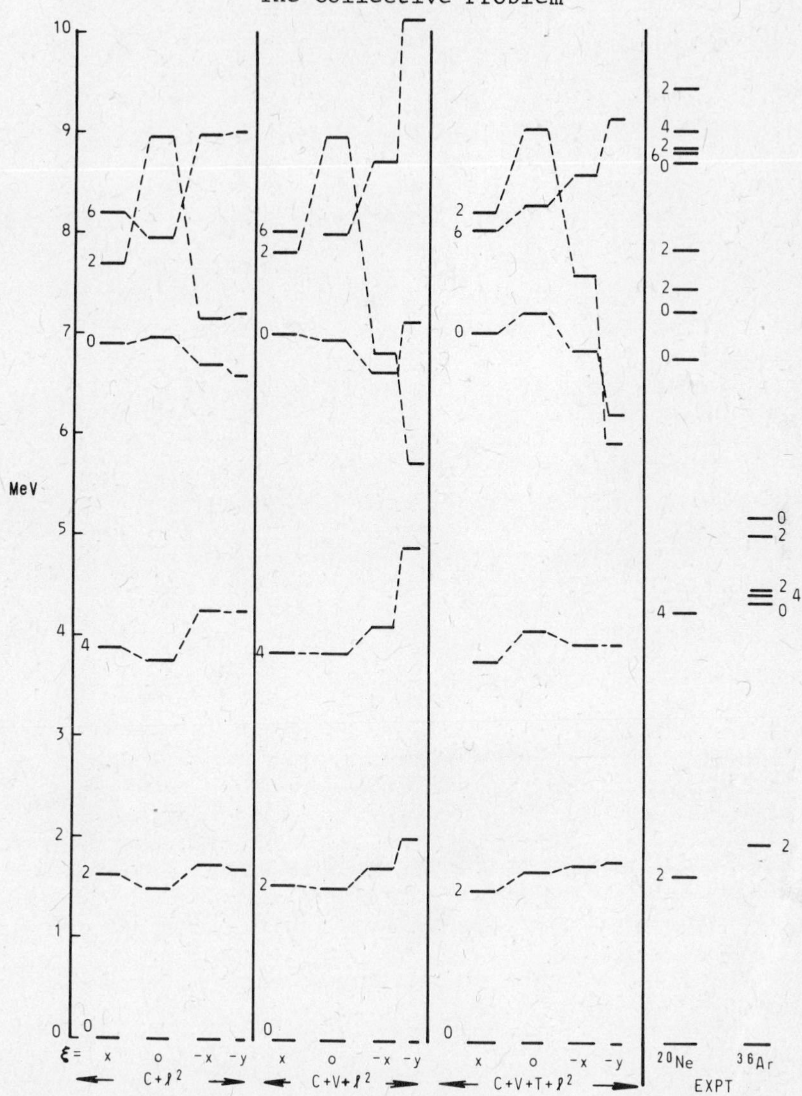

Fig. 6. Effect of spin dependent forces in the four fermion system in the shell model calculation. In each set of four spectra the spin orbit force is varied among x = 2 MeV (appropriate to Ne^{20}), 0, -x, and +y = -2.5 (appropriate to Ar^{36}). The set of spectra on the LHS were computed with the two body matrix elements of the central (C) force alone; for the middle set the two body spin-vector force has been added to the central (C+V); and for the RHS set the full Kuo potential including the tensor face has been used (C+V+T). The s-d splitting (ℓ^2) was set at 0.8 MeV. The positive parity low energy spectra for Ne^{20} and Ar^{36} is shown on the right. (In collaboration with S.S.M. Wong.)

phonon spectrum we seek degeneracy of a 0^+, 2^+ and 4^+ triplet at about twice the energy of the first excited $J=2^+$ state. We note that the energy of the (second) $J=2^+$ state is very sensitive to the spin orbit force especially with the negative coefficient x at the end of the shell. We also note that for negative x both the spin orbit and two-body spin vector lower the 2^+ state whereas for positive x these two forces seem to oppose one another. Unfortunately, the 2^+ is now lowered nearly enough to form any triplet. It is interesting to note that in both Ne^{20} and Ar^{36} there are two close lying 0^+ states and 2^+ states in the excitation spectrum (at about 7 MeV) in Ne^{20} and about 5 MeV in Ar^{36}). It is well known that in Ne^{20} one state in each pair is a core excited state. The question now arises whether one state in each pair in Ar^{36} is also a core excited state? If this is so then are the core excited states part of the triplet? If they are part of the triplet, then is the triplet really a two-phonon triplet i.e. carrying enhanced B(E2) to the one-phonon state? As can be seen from Table 1, the 2^+ state in the

TABLE I

The effect of the two body spin-vector and spin-tensor forces on the B(E2) transitions between low energy states in Ar^{36} in Weisskopf units. (N.B. The small $B(E2:2_2 \rightarrow 2_1)$ is indicative of a small two-phonon structure for the 2_2 state. In collaboration with S.S.M. Wong.)

$B(E2:j_i\ J_f)$ for Ar^{36} (Weisskopf Units)

J_i	J_f	C	C + V + T
2_1	0_1	16	16.95
0_2	2_1	6.7	5.08
4_1	2_1	20.2	20.61
2_2	2_1	2.3	1.46
2_2	0_1	0.03	.06

shell model does not pick up any B(E2) strength as it is lowered by the spin dependent forces, thus providing an argument against its collective nature. If the core excited states are necessary to form the second phonon triplet then the calculation of the low energy spectrum would not separate neatly into shell model and core excited states as in Ne^{20} but would be very complex. Ar^{36} is thus a very interesting nucleus at this time and more experimental evidence for its vibrational structure would be most welcome.

At the moment, unfortunately, we are not able to do a similar analysis for S^{32} where the collective vibrational structure seems to have been established. In order to carry the shell model calculation into the shell we need some reliable way of truncating the basis. If the Q.Q. force dominates the residual interaction then it is natural to select states that are classified according to SU_3. The question now is what effect does the spin orbit force have on this scheme? A couple of years ago, with Takashi Sebe, we proposed a way of analyzing the effect

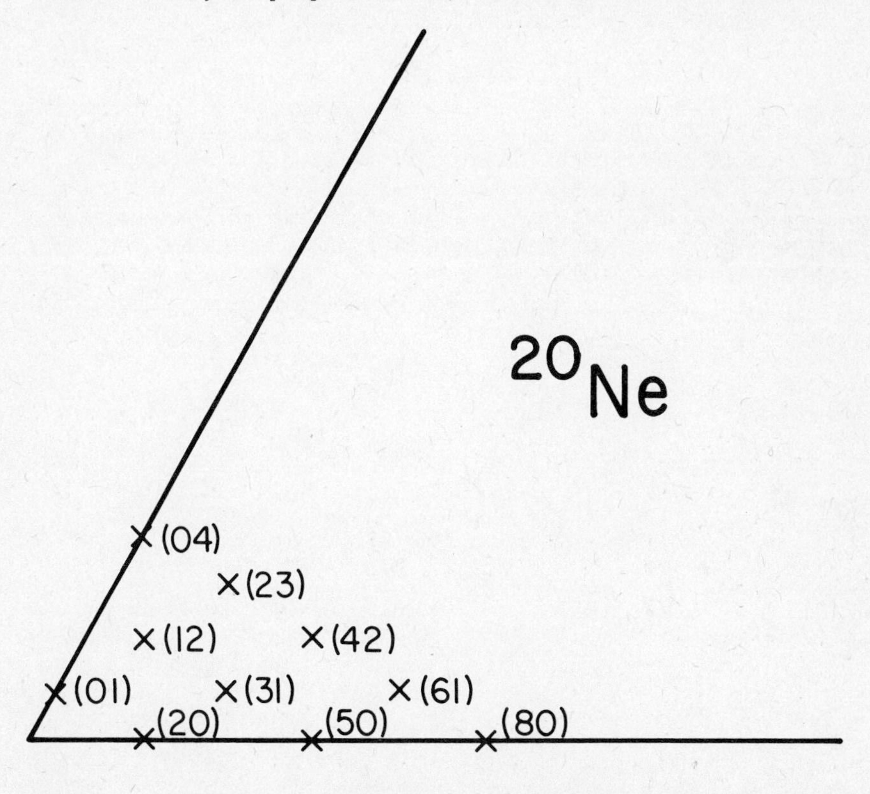

Fig. 7. Places on the βγ-plane of allowed representations of SU_3 for Ne^{20} (*c.f.* Ref. 17).

of the spin-orbit force in the SU_3 model that suggested the retention of collective features[17]. We noted that the representation labels ($\lambda\mu$) of the group SU_3 define the $\beta\gamma$-deformation of the "intrinsic-state" underlying the SU_3-band. Now, for a given number of particles in the shell one can mark the position on a $\beta\gamma$-plane of the allowed ($\lambda\mu$)-representations of SU_3. Figure 7 shows such a drawing for Ne^{20}. If the amplitude of a given representation in the lowest eigenfunction is represented by the side of a box encompassing the appropriate point on the $\beta\gamma$-plane, then the full function takes on the pictorial form shown in Figure 8. This is a recent calculation on the mass 21 system by Sebe for a $20x \sum_{ij} Q_iQ_j + 5(1-10x) \sum_i (s_i\ell_i)$ potential. Shown in the figure are the structures of the two lowest J = 5/2 states for two different strengths of the Q.Q. force. We assign K- labels to the states because they have similar structures to other states of different spin - for example the K = 1/2 J = 5/2 has a similar pattern to that shown in Figure 9 for the K = 1/2 J = 1/2. This particular band has a very simple structure reminiscent of

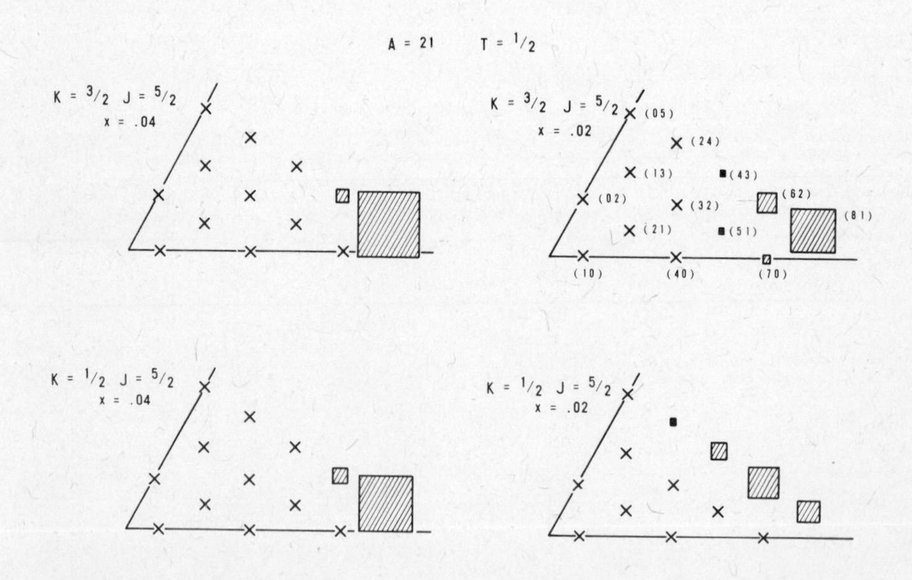

Fig. 8. Pictorial representation of the eigenfunctions of the Hamiltonian $H = 2x \sum_{ij} Q_iQ_j - 5(1-x) \sum_i s_i\ell_i$ for two lowest J = 5/2, T = 1/2 states of mass A = 5 system. The areas of the boxes are proportional to the probability of finding the eigenfunction of the SU_3 representation of that position in the $\beta\gamma$-plane. (In collaboration with T. Sebe.)

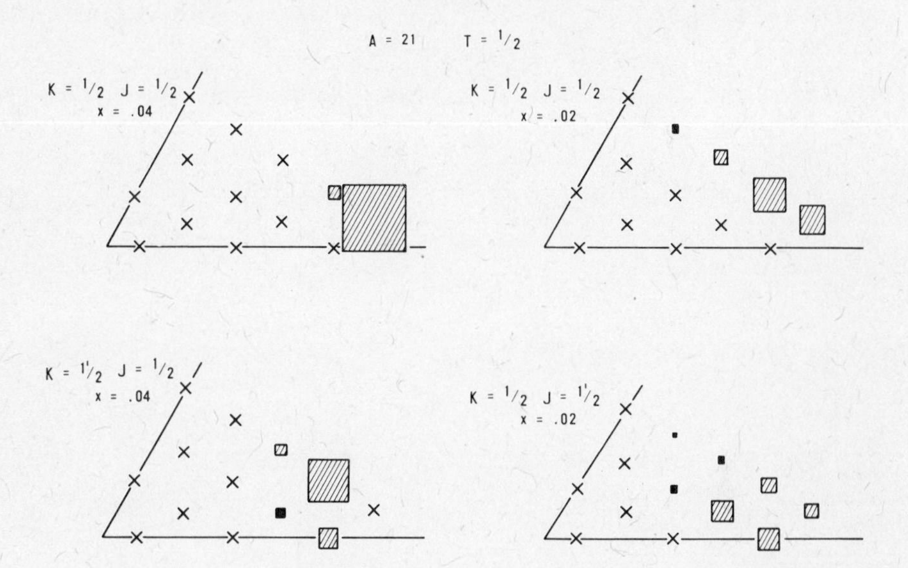

Fig. 9. As in Fig. 8 but for the J = 1/2, T = 1/2 states.

a γ-vibrational structure - and similar to the type of admixtures found[17] in Ne^{20}. It was this simplicity that suggested that even with spin dependent forces the SU_3-scheme is still very useful and could lead to a realistic way of truncating basis. The second K = 1/2 band does not appear to have such a simple structure and, in the sense of the SU_3 coupling scheme needs many configurations to build up the wave function. We had hoped that the simple admixture of representations would persist for low energy states for all nuclei but this does not appear to be so. In Figure 10 we show results for the mass 22 system where it is the lowest J=1, T=0 state that has a complex structure with the spin orbit force: the second J=1, T=0 state has the simple "γ-vibrational-like" structure again. This particular calculation is slightly disappointing from the point of view of finding a good truncation scheme.

It is tempting to try to associate the "γ-vibration-like" structure in the mixed SU_3-representations with the "softening" of the average single particle self-consistent field. We know that the Hartree-Fock states arising from a Q.Q.-force is almost exactly the intrinsic state of the "leading" representation[3] of SU_3. Actually this intrinsic state is apparently generated by the Hartree-Fock procedure for any central force. In Table II, for example, we show the overlap in column 4 of the intrinsic state of the (80)-representation for Ne^{20} with the results of a Hartree-

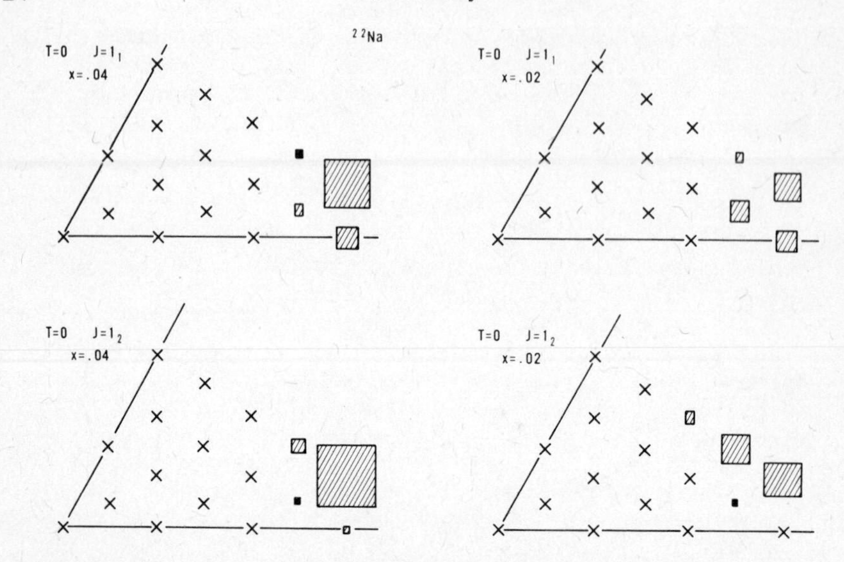

Fig. 10. As in Fig. 8 but for the J = 1, T = 0 states of the mass A = 22 system.

Fock calculation for a central + spin orbit force (strength-x). The central force was that derived from the Kuo potential[15]. We note that for x = 0 the overlap with the SU_3 state is almost complete. In the last column we show the overlap of the above Hartree-Fock states with the Hartree-Fock state arising from the full Kuo potential with spin orbit strength x = 2 MeV. We note that the two-body vector and tensor forces do not contribute significantly to the average field. The main difference then between Hartree-Fock calculations and the pure SU_3-model is in the single particle spin-orbit force.

But does the spin orbit force merely modify the average field or does it also effect the stability with respect to vibrations - and which is really the dominant effect? Can we correlate the deviation away from the SU_3 structure with the addition of spin-orbit force in the Hartree-Fock model with that seen in the shell model calculations? With Axel Jensen, we have examined this problem by constructing the energy surfaces in the $\beta\gamma$-plane around the Hartree-Fock minimum for all even-even nuclei in the 2s-1d shell. We tackle the problem by solving the constrained Hamiltonian

$$H' = H - \lambda Q_o^{\,2} - \mu\,(Q_2^{\,2} + Q_{-2}^{\,2})$$

for quadrupole operators $Q_\mu^{\,2} \sim r^2 Y_\mu^{\,2}$ and various λ and μ.

Such a procedure cannot map the whole energy surface but only that in the immediate vicinity of a minimum - in fact up to the "turning-point" of a curve through the minimum. As it turned out, the turning-points were very

TABLE II

Overlap for mass A = 4 system of the Hartree-Fock state derived from a Hamiltonian H = central + $x \sum_i x_i \ell_i$ (for various values of x) with (column 4) the intrinsic state of the leading $(\lambda\mu) = (80)$ representation of SU_3 and (column 5) the Hartree-Fock state derived from the full Kuo potential with spin-orbit parameter x = 2 MeV. Columns 2 and 3 give the energy and quadrupole moment of the original Hartree-Fock problem as a function of spin-orbit strength x. (N.B. For negative x the calculation applies to the 4-hole system, i.e. Ar^{36}.) (In collaboration with A. Jensen.)

Ne^{20}

$-E_T = 24.713$, $Q_T = 15.45$

x	-E	Q_o	OV(SU_3)	OV(Full)
-4	32.127	11.02	0.40	
-3.5	29.821	12.41		0.35
-3	27.876	13.57	0.67	0.47
-2	25.068	15.05	0.87	0.69
-1	23.529	15.72	0.96	0.84
0	23.055	15.91	0.99	0.94
1	23.483	15.76	0.97	0.99
2	24.687	15.33	0.91	1.00
3	26.562	14.66	0.82	0.97
4	29.002	13.81	0.72	0.92

close to the minima and so very little of the energy surface could be drawn. In Figure 11 we show the right-hand side of the $\beta\gamma$ plane of Figure 8 for the mass 4 system for a central + spin orbit force and various strengths for the spin orbit force. The unmarked crosses (x) mark the position of the closest representation of SU_3 (leading representation at 16) and the cross labelled with an M denotes the position of the absolute minimum. We note the following points:

(a) The procedure fails to map a surface which would be large enough to encompass many SU_3 representations but it does tend to indicate a softening in the γ-direction.

(b) The surface is flatter for increasing strength of the spin orbit force especially for negative values corresponding to *holes* in the shell.

The failure to draw much of the surface was disappointing as we did have hopes that such a procedure

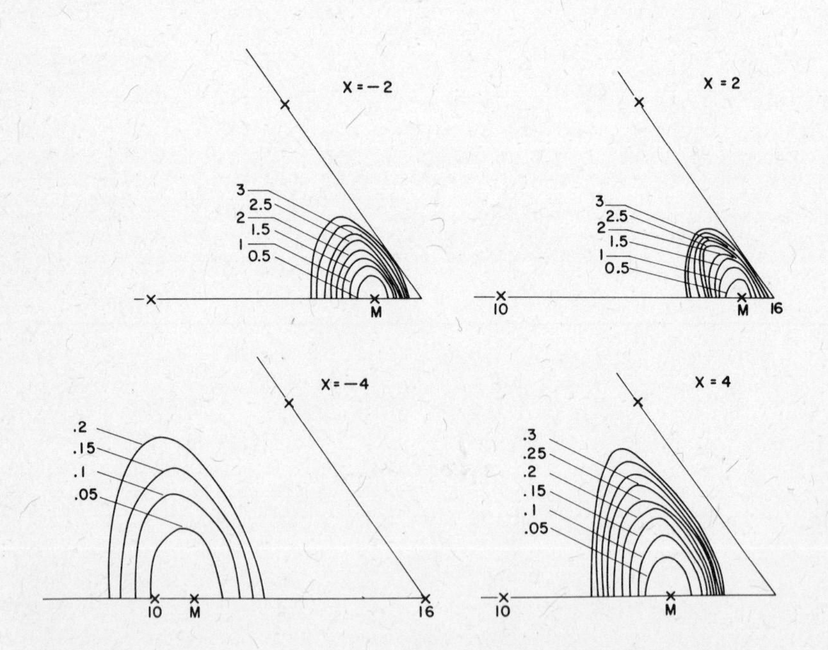

Fig. 11. Part of the energy surface for large β for the mass A = 4 system near the absolute Hartree-Fock minimum (H); shown for various values of the spin orbit strength x. The unlabelled crosses correspond to the nearest representations of SU_3 (($\lambda\mu$) = (80) is at the point marked 16). (In collaboration with A. Jensen.)

would tell us which SU_3 representations were important. The actual flattening of the energy surface is interesting but in itself tells us very little about vibrational features until the mass parameters for vibrations in a $\beta\gamma$-plane have been computed. Using formulae of Sobiczewski *et al.* we have computed these mass parameters and hence derived the minimum harmonic vibrational frequency $\hbar\omega$ in the $\beta\gamma$-plane as a function of the strength of the spin-orbit force. Table III shows the results for the mass 4 system; x=2 corresponds to Ne^{20} where it is seen that the minimum vibrational frequency prefers the γ-direction (90°) consistent with the SU_3 representation mixing in the shell model; x = -2 corresponds to the holes in Ar^{36} and the results can be interpreted as the particles in an oblate shape vibrating in the β-direction. The actual vibrational frequency is unfortunately very close for x = ±2 thus providing little fuel for our idea that the spin orbit force is responsible for the vibrations. Note that the frequency $\hbar\omega$ decreases as spin orbit strength increases. In Table IV we show the results for the oblate solution in Si^{28} which tends to want to vibrate in the γ-direction and in Table V the results for the prolate solution in Si^{28} which wants to vibrate in the γ-direction. Surprisingly the vibrational frequency for Si^{28} and Ne^{20} are about the same (8 to 9 MeV) despite the fact that Si^{28} has a lower 0^* excited state than Ne^{20}. One must presume that the harmonic approximation implicit in the formulae of Sobiczewski *et al.* is not adequate to describe the vibrational effects in Si^{28}. Thus the spin-effects might be important in the vibrational phenomena but we are using the wrong formulae.

What can we conclude from these calculations?

1. Rotational features at the beginning of the 2s-1d shell are in the spirit of the SU_3 shell-model, i.e., we should expect band cut-offs.

2. Vibrational spectra do exist at the end of the shell but it is not obvious that the shell model is capable of reproducing them.

3. The two-body vector and tensor forces do not contribute much to the average Hartree-Fock field - and what they do contribute can be simulated by an additional single particle spin-orbit force.

4. The single particle spin-orbit force describes the main differences between the SU_3 model and the Hartree-Fock model.

5. The spin-orbit force tends to soften the nucleus

TABLE III

Harmonic vibrational frequencies according to formulae of Ref. 18 for the mass A = 4 (e.g Ne^{20}) system as a function of spin-orbit strength x. There are two orthogonal normal modes in the $\beta\gamma$-plane; columns 2 and 5 give respectively the maximum frequency and the direction of vibration (measured from the line $\gamma = 0$; columns 3 and 6 are similar to columns 2 and 5 but for the minimum frequency. Column 4 shows the energy gap. N.B. Negative x corresponds to holes in the 2s-1d shell. (In collaboration with A. Jensen.)

x	$(\hbar\omega)_{max}$	$(\hbar\omega)_{min}$	Gap	$(angle)_{max}$	$(angle)_{min}$
-4	7.71	5.25	4.89	0^0	90^0
-3	7.65	7.23	6.75	0^0	90^0
-2	8.76	8.14	8.01	90^0	0^0
-1					
0	12.64	10.88	9.17	90^0	0^0
1	9.64	9.78	8.48	0^0	90^0
2	9.51	8.17	7.63	0^0	90^0
3	9.54	7.18	6.77	0^0	90^0
4	9.70	6.32	5.95	0^0	90^0

TABLE IV

As in Table III but for the oblate solution of the mass 12 system (e.g. Si^{28}).

x	$(\hbar\omega)_{max}$	$(\hbar\omega)_{min}$	Gap	$(angle)_{max}$	$(angle)_{min}$
-4	7.45	3.63	3.43	60^0	150^0
-3	8.75	5.97	5.67	60^0	150^0
-2	10.3	8.17	7.83	60^0	150^0
-1	11.6	10.3	9.86	60^0	150^0
0	15.68	14.79	10.70	150^0	60^0
1	11.39	10.51	10.19	150^0	60^0
2	10.77	9.29	9.00	150^0	60^0
3	10.20	8.36	8.10	150^0	60^0
4	9.79	7.91	6.61	150^0	60^0

TABLE V

As in Table IV but for the prolate solution of the mass 12 system (e.g. Si^{28}).

x	$(\hbar\omega)_{max}$	$(\hbar\omega)_{min}$	Gap	$(angle)_{max}$	$(angle)_{min}$
-4	9.77	7.79	6.65	90^0	0^0
-3	10.18	8.38	8.09	90^0	0^0
-2	10.75	9.30	9.00	90^0	0^0
-1	11.28	10.35	10.01	90^0	0^0
0	15.78	14.63	10.47	90^0	0^0
1	10.95	10.08	9.78	0^0	90^0
2	9.83	8.17	7.87	0^0	90^0
3	8.45	6.04	5.72	0^0	90^0
4	7.28	3.68	3.45	0^0	90^0

although, in the harmonic approximation this appears to be no greater for nuclei at the end of the shell than at the beginning.

It would be nice to have more data on Ar^{36} to see whether it is a good vibrator or not and hence to test the adequacy of the shell model.

ACKNOWLEDGMENTS

I wish to thank my collaborators, B. Castel, K. Stewart, S.S.M. Wong, T. Sebe, and A. Jensen, for their invaluable assistance and discussions. Thanks are also extended to my experimental colleagues in Chalk River, Queens University and Toronto University for permission to quote unpublished data.

REFERENCES

1. H.E. Gove, *Proceedings of International Conference on Nuclear Structure*, University of Toronto Press, Kingston, 1960.

2. J.P.Elliott, Proc. Roy. Soc. A245, 128 (1958), and A245, 562 (1958).

3. M. Harvey, *Advances in Nuclear Physics*, Vol 1, ed. M. Baranger and E.W. Vogt, Plenum Press, New York, 1968.

4. A. Bohr, Phys. Rev. 81, 134 (1951).

   A. Bohr, Dan. Mat. Fys. Medd 26, No. 14 (1952).

   A. Bohr and B. R. Mottelson, Dan. Mat. Fyps. Medd 27, No. 16 (1953).

5. See Ref. 3, Appendix G, Equation G.25.

6. C.S. Kalman, J.P. Bernier and M. Harvey, Can. Jour. of Phys. 45, 1297 (1967).

7. A.E. Litherland, Panel Discussion page 618, *International Conference on Properties of Nuclear States*, University of Montreal Press, Montreal, 1969.

8. O. Hausser, T.K. Alexander, A. McDonald, A.E. Litherland and W. Diamond (Private Communication).

9. F. Ingebretsen, B.W. Sargent, A.J. Ferguson, J.R. Leslie, A. Henrikson, and J.H.Montague (Private Communication).

10. K. Nakai, J.L. Quebert, F.S. Stephens and R.M. Diamond, Phys. Rev. Lett. 24, 903 (1970).

11. O. Hausser, T.K. Alexander, A. McDonald, W. Diamond (Private Communication).

12. B. Castel, K. Stewart and M. Harvey, Nucl. Phys. to be published.

13. V.K. Thankappan and S.P. Pandaya, Nucl. Phys. 19, 303 (1960), and V.K. Thankappan, Phys. Lett. 2, 122 (1962).

14. S.S.M. Wong and M. Harvey, Contribution 7.35, *International Conference on Properties of Nuclear States*, University of Montreal Press, Montreal, 1969.

15. T.T.S. Kuo, Nucl. Phys. A103, 71 (1967).

16. J.B. French, E.C. Halbert, J.B. McGrory and S.S.M. Wong, *Advances in Nuclear Physics*, Vol. 3 ed. M. Baranger and E.W. Vogt, Plenum Press, New York 1969.

17. M. Harvey and T. Sebe, Nucl. Phys. A136, 459 (1969).

18. A. Sobiczewski, Z. Szymanski, S. Wyceck, S.G. Nilsson, J.R. Mix, C.F. Tsang, C. Gustafson, P. Moeller and B. Nilsson, Nucl. Phys. A131, 67 (1969).

19. O. Hausser, B.W. Hooton, D. Pelte, T.K. Alexander and H.C. Evans, Phys. Rev. Lett. 22, 359 (1969)

DISCUSSION

ZAMICK: In most heavy nuclei, it seems that the J=0, two phonon state is higher than the J=4 and the J=2 (two phonon state), but in this case it is consistently lower. In your case, do you know why that was so?

HARVEY: I really don't know. We have been trying to calculate the anharmonic effects in S^{32}, however, most people, when they describe vibrations, start out in an independent particle model, build up to the ground state and say the phonon state is a particle-hole and two-particle two-hole admixture. Now with Castel and Stewart we try to start out at the other extreme by saying that we have a phonon state and then try to describe the particle-hole and two-particle two-hole deviations from the real state as we try to describe the anharmonic effects in the particle-hole formalism. The results are quite encouraging except that for S^{32} the quadrupole moment of the first 2+ state is too small. Also, the 0+ state, that Larry Zamick was talking about, stays above the 4+ state.

WILDENTHAL: I dislike quoting other people's data from memory, but Garvey and Mermaz, at Oxford, have looked at Ar^{36} and speaking qualitatively the B(E2)'s do not exhibit the behavior that is seen in S^{32}. The transitions to the 2+ from the J = 0+,2+,4+ multiplet are not factors of two and three stronger than that from the J = 2+ to 0+.

HARVEY: It is not a vibrator?

WILDENTHAL: In that sense you spoke of the extra J=2 and J=0 states perhaps being core excited, but perhaps excitations into the 7/2 shells may be more germane.

HARVEY: I am sorry, but I meant to include that possibility when I spoke of core excited particles.

SEGEL: Do you make any predictions from the B(E2)'s from the second 2+ state to ground? [In Ar^{36}].

HARVEY: In our calculation it was very small. The value is given on Table I.

SEGEL: I think that is the one case that hasn't been measured. Usually they are about a single particle unit.

HARVEY: This is two orders of magnitude less in the calculations than the two phonon to the first phonon, which was about one Weisskopf unit. The one phonon ground was about an order of magnitude greater than that.

SHARMA: I would like to know how to predict band cut-off on the basis of the SU_3 model? In the case of many nuclear reactions, the high spin states in the rotational bands of daughter nuclei have decreasing relative intensity, as found from cross section measurements (as in the case of $Tb^{159}(d,t)Tb^{158}$). The weaker relative intensities for higher spin states are an indication of expected band cut-off. Is there any prediction of this type on the basis of theory?

HARVEY: The cut-off is really a shell model feature. For instance, if I put four particles in the 2s-1d shell, I can't construct a state of spin 10.

SHARMA: Actually there is some core contribution also, and the angular momentum of the nuclear core is not limited. How can you say that there is any band cut-off?

HARVEY: Strictly speaking, we cannot. As the angular momentum builds up you can say, OK, then, the core stretches so that you can actually build up a rigid rotor. That's why the experiment on Ne^{20} was very interesting because it tells us as we go up the band whether the core is really deforming and whether we are really developing a rigid rotor or whether in fact there is enough shell structure there to prevent this developing. Now unfortunately, the finding of the spin 10 state doesn't tell us that it is part of a rotational band until we see an enhanced $B(E2: 10 \rightarrow 8)$ transition: and this is very hard experimentally. That is why the experiments have been following the B(E2)'s going up the band and looking at the trends, that is, to see if they go on the trends of the rotational model or on the trends of the shell model. They seem to follow the shell model.

CASTEL: Isn't there a difficulty in using the Hartree-Fock calculations to investigate the B and C in the second half of the 2s-1d shell? There, you really want to investigate the vibrational structure and we know that you obtain the oblate solution for S^{32} if you don't have enough spin-orbit although the prolate one is observed.

HARVEY: What we were trying to see is that as we turn on the spin-orbit force the vibrational states for the holes come down so low in energy that you would say it is really pointless considering a stable Hartree-Fock state even though there is a Hartree-Fock minimum. The low energy of the vibrational states would imply that the Hartree-Fock field is so soft that one shouldn't take it. That is what we were trying to prove, but unfortunately, it didn't come out of the calculation; the vibrational state was still high in energy for the hole configuration.

PROSSER: This may be related to Dr. Zamick's question. I noticed in the S^{32} case that the second 5/2 state remained consistently above the 7/2 state, independent of the quadrupole moment. I wonder if you have any explanation for that since this is not what is seen experimentally?

CASTEL: I think that the main contribution in the 7/2+ state comes from the coupling of the 3/2+ state to the second phonon state and there is very little anomalous structure in the second phonon states.

PROSSER: There is apparently some other term that you have left out that is necessary to invert the order of these states?

CASTEL: Yes, they would come from quadrupole moments of the second phonon state.

WILDENTHAL: One more general question now. It really does seem that perhaps S^{32} is unlike any of the even-even nuclei above it or below it. I am curious as to why this is so? The shell model results are that you don't get this enhancement of a factor of two or three from the second J=0,2,4, down to the J=2 and experimentally you only get this enhancement in S^{32} - you don't get it in S^{34} or, I think, Ar^{36}. S^{32} looks rather anomalous.

I.a. ON NEGATIVE PARITY STATES IN ODD-A NUCLEI

B. Castel
Queen's University
Kingston, Canada

We have seen from Malcolm Harvey's discussion that the systematics of even parity states in odd-A nuclei in the second half of the 2s-1d shell can be studied in terms of a core-quasi particle coupling model. In the same region, negative parity states are observed at about the same energy in various odd-A nuclei as we see in Figure 1. It is then interesting to see whether the behaviour of these low-lying negative parity states can be understood in an extension of this core-quasi particle model just discussed.

The decay from these negative parity states, presumably dominated by $f_{7/2}$ and $p_{3/2}$ single particle character presents interesting features. For instance in Cl^{35}, it has been observed that all the E1 transitions originating from the $7/2^-$ state are very strongly inhibited. A factor of 10^{-8} is observed for the E1 transition between the $7/2^-$ and first $5/2^+$ state.[1,2]

It is clear that these extreme retardations are rather difficult to account for; especially if the only negative parity states taken into consideration are single particle ones. We note however, from Figure 2 that in this region of the 2s-1d shell, all even-even nuclei exhibit a low-lying 3^- octupole state, and in the case of S^{32} and Ar^{36} for instance, the enhanced E3 ground state decay is seen together with a retarded E1 transition to the first 2^+ state.[3] The coupling of this octupole phonon state to the 2s-1d shell single quasi-particle states may therefore be expected to contribute significantly to the wave function of the negative parity states in neighbouring odd-mass nuclei. In the quasi-particle-core coupling model, a description of negative parity states involves the additional consideration of the octupole-octupole type of interaction and the $7/2^-$ state for instance would then be described as a purely single-particle part, a quadrupole part and an octupole one:

$$|7/2^-\rangle = \alpha|7/2^-, 00\rangle + \beta|7/2^-, 12\rangle + \gamma|3/2^+, 13^-\rangle + \ldots$$

$$|5/2^+\rangle = \alpha'|5/2^+, 00\rangle + \beta'|5/2^+, 12\rangle + \gamma'|3/2^+, 12\rangle + \ldots$$

For comparison, we have shown the principal components of the first $5/2^+$ state, a typical positive parity state. Here only the main components of the wave functions are given. In the single particle part of the contribution to the E1 rate (indicated by thin arrows) a large inhibition factor arises because the $d_{5/2}$ subshell is nearly filled while the $7/2^-$ orbitals are still little occupied, thus in the absence of any collective behaviour, this would lead to an inhibition factor of typically 10^{-4} in the E1 rates. It is known, however, as we mentioned before that in S^{32}, for instance, the collective contribution would also be of this order. In the case of the $7/2^- - 5/2^+$ transition the calculated and collective terms almost cancel to give a further substantial inhibition, as we see in Table I.

In the E3 part of the ground state decay, both the single particle and collective parts contribute to an enhanced transition strength which is, in fact, consistent with what has been observed experimentally. It is therefore

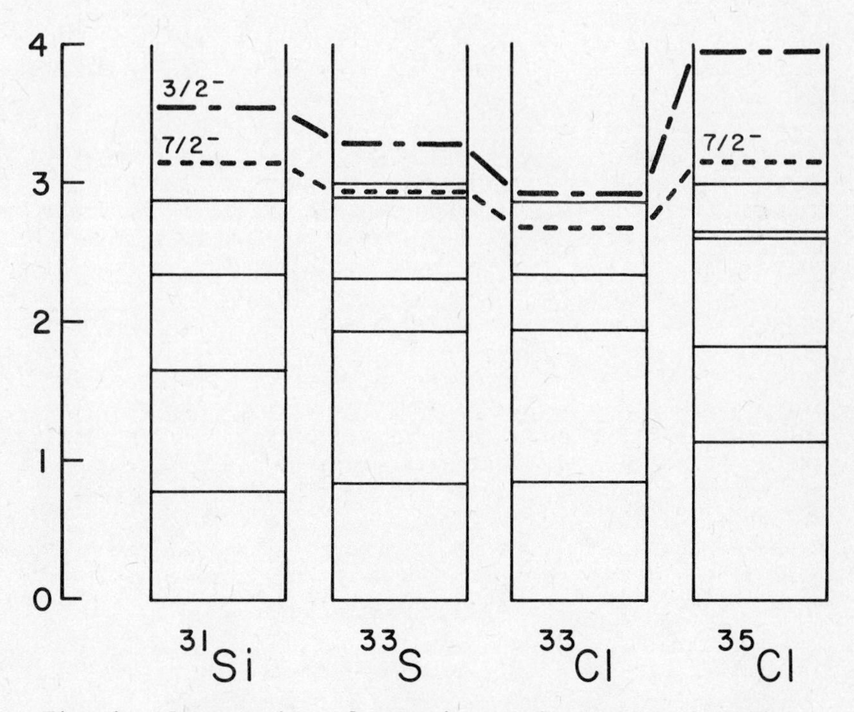

Fig. 1. Systematics of negative parity states in odd-A nuclei with A = 31, 33, 35, and 37.

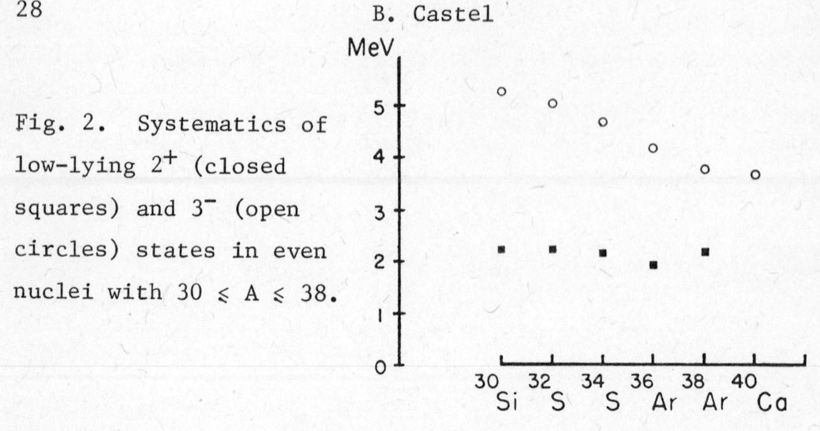

Fig. 2. Systematics of low-lying 2^+ (closed squares) and 3^- (open circles) states in even nuclei with $30 \leqslant A \leqslant 38$.

interesting to notice that a relatively simple model can account for the decay properties of the $7/2^-$ state in Cl^{33}. More should be known on the decay rates of other negative parity states in that region to investigate the possibilities of this simple model.

This talk describes work carried out by Dr. K.W.C. Stewart and myself. It was supported by the National Research Council and the Atomic Energy Board of Canada.

Table I

E1,E3 Rates in Cl^{35} (W.u.)

Transition	Calc.	Exp.[1]	Exp.[2]
$B(E1: 7/2^- \rightarrow 5/2^+)$	1.2×10^{-7}	$\leqslant 2 \times 10^{-8}$	$\leqslant 5 \times 10^{-8}$
$B(E1: 7/2^- \rightarrow 7/2^+)$	3.4×10^{-3}	$\leqslant 10^{-5}$	$\leqslant 10^{-5}$
$B(E3: 7/2^- \rightarrow 3/2^+)$	2.1	2.8	6.5

REFERENCES

1. F. Ingebretsen, T.K. Alexander, O. Haussen and D. Pelte, Can. J. Phys. 47, 1295 (1969).

2. P. Taras and J. Matas, Can. J. Phys. 48, 603 (1970).

3. F. Ingebretsen, B.W. Sargent, A. J. Ferguson, J.R. Leslie and J.H. Montague, Nucl. Phys., to be published.

J.P. Thibaud, M.M. Aleonard, D. Castera, Ph. Hubert, F. Leccia and P. Mennarth, Journal de Physique 31, 131 (1970).

4. B. Castel, K.W.C. Stewart and M. Harvey, to be published.

DISCUSSION

PROSSER: Gale Harris and I have recently remeasured the decay of the $7/2^-$ analog state in Cl^{35} and among other things have new information on the decay of the 3.16-MeV state. We have seen the decay from it to the $5/2^+$ state and, if it were pure E1, it would be about $1.6 \pm 0.2 \times 10^{-8}$ Weisskopf units, again assuming pure E1, and it also decays with a 160-keV transition to the next $5/2^+$, at 3 MeV, with a strength of $1.5 \pm 0.3 \times 10^{-4}$ Weisskopf units. So this is a much less inhibited E1. In doing this we took measurements at 0^0 and 90^0. These are very weak transitions, obviously, so I don't want to make strong claims about mixing ratios. However, the anisotropy observed for the decays to the 2.6 MeV and the 3 MeV states agree with pure E1. On the other hand, the decay to the lower 5/2 at 1.7 MeV does not show the same anisotropy, but one which can be explained by an M2 admixture with a mixing ratio on the order of 0.5. This would mean that the E1 is even weaker and that the M2 is quite standard. So that this rather bears out what you had on your slide and gives some precise numbers rather than limits.

CASTEL: Do you have any data on the E3?

PROSSER: On which one, the one from the 3.16 MeV state? [Yes]. There is essentially no change in that number from that that has been published. We have seen the cross-over from the compound state to ground. This agrees with pure M2 and is about three Weisskopf units. This will be reported at Houston by Dr. Harris. [Bull. Am. Phys. Soc. 15, 11 (1970), paper CB6.].

I.B. ISOSPIN RESONANCES IN 4N NUCLEI[+]

Stanley S. Hanna
Department of Physics, Stanford University
Stanford, California

I. INTRODUCTION

Two years ago at this conference I made a very brief report on the first observation of the isospin forbidden T=2 resonances.[1] Since there has been a lot of activity in this area since that time, I thought today it would be appropriate to look at the general picture of the T = 2 resonances and see where we stand now.

The T = 2 resonances in which we are interested and on which almost all the work has been done are in the self-conjugate nuclei and principally in the 4N nuclei. The attractive feature of these resonances is that they fall in a region of excitation energy where the proton and alpha channels are open energetically but closed by isospin conservation, so that if isospin is well conserved you might not be able to observe them by resonance reactions at all. However, if there is some kind of mixing or, of more interest, a breakdown of true isospin conservation, then you might be able to observe these resonances with particle reactions. The case of mixing would be of interest because you would observe the resonances through the admixture of the isospin allowed part of the wave function.

Now the T = 2 levels fall above the T = 1 and the T = 0 levels in these nuclei and there is a basic selection rule for electromagnetic radiation which says that you may not change the total isospin in an electromagnetic transition by more than one unit, so you have the interesting fact that you may see isospin cascades, i.e., gamma transitions from the T = 2 to the T = 1 levels down to the T = 0 levels. Thus, in studying the T = 2 levels you are involved immediately in the study of the T = 1 levels. They are interesting, of course, in their own right. The T = 1 levels in these nuclei occur at an excitation energy where the neutron and proton channels are energetically closed, but in almost all the cases the alpha channel is energetically open. Again, in this case, the alpha channel is closed by isospin conservation, so that these levels are interesting to study to see how large the alpha width is in them.

[+]Supported in part by the National Science Foundation.

In addition, the lowest lying T = 2 and T = 1 levels in these nuclei are thought to have rather simple configurations. This can be shown both experimentally and theoretically. Experimentally, the T = 2 levels in these nuclei in general are formed, in fact this is the way in which they have all been first observed, by very strong 2-particle transfer reactions, so that basically they are 2-particle, 2-hole states in these 4N nuclei. This is also borne out by the fact that you can reach them with two very strong M1 excitations, i.e., starting from the ground state of the nucleus you can excite the lowest 1^+, T = 1 level by a strong M1 excitation and then you can excite the 0^+, T = 2 level by another strong M1 excitation, showing again the basic 2-particle character of these states.

So, first of all I would like to look at the experimental situation which has developed over the past two years and then, insofar as time permits, I would like to discuss some of the interesting points regarding the nature and properties of these states.

II. THE T = 2 RESONANCES

Figure 1 is a summary of T = 2 levels which have been observed in the A = 4N nuclei. Along with them, some of the T = 1 levels are shown, at least those which are involved in the decay of the T = 2 levels. Now this is a highly schematic diagram. It is meant to indicate which nuclei have been studied, what states have been found, and if gamma decays have been observed. It is not meant to show all the gamma decays, nor all the states or the fine structure, but just to show the general picture. This figure was prepared about a year ago, but in its overall features it is still essentially correct. There have been a few refinements which I want to discuss, but first I want to describe the figure.

The T = 2 levels are always at the top in each nucleus. As you can see, they have been observed in 4N nuclei all the way up to mass 56 with the exception of mass 48. It is perhaps an unfortunate notation, but the fuzzy notation is meant simply to indicate that the state was found in a 2-particle transfer reaction and therefore its energy is not well known, but not that the state is not sharp. We believe in most cases that the states are sharp, of the order of 10 keV or less, but that in these cases the energies are not too well known because of the reaction that was used. In the other cases, shown as sharp lines, the levels have been observed as resonances in scattering

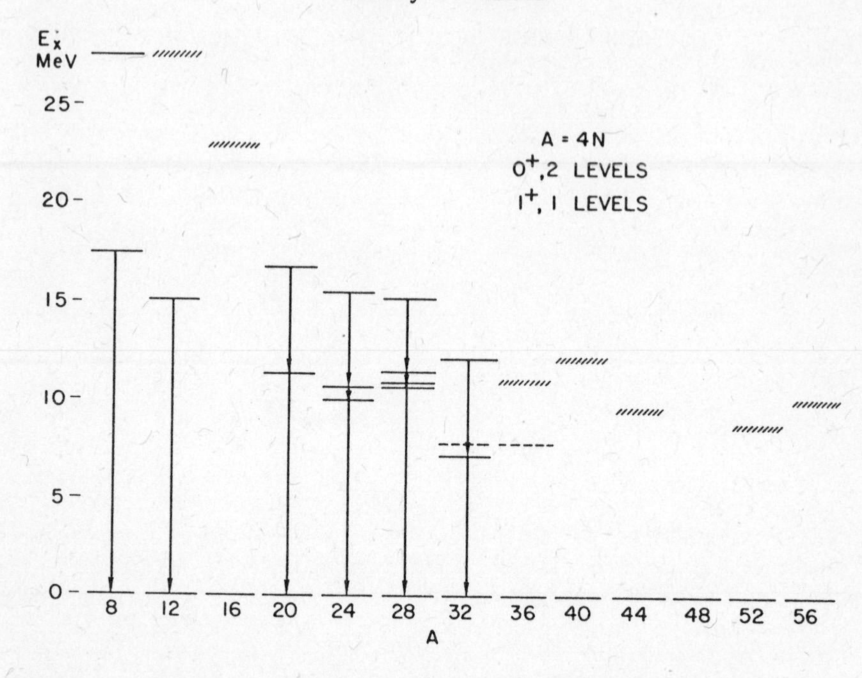

Fig. 1. Summary of T = 2, 0^+ levels (at the top) in 4N Nuclei and of T = 1, 1^+ levels (in the middle) to which they are connected by strong M1 transitions. The hatched lines indicate levels located by transfer reactions whose positions are not accurately known. The other T = 2 levels are located by resonance reactions. The T = 2 levels in C^{12} and O^{16} have recently been seen in resonance reactions (see text). The arrows indicate symbolically that the major gamma transitions have been studied, but all transitions are not shown.

or capture and therefore, in principle, the energies are well known, usually to about 5 keV or less.

Now, if you are able to form the resonance in a compound nucleus reaction, you always have the possibility of studying the gamma decay. The gamma transitions have been studied in at least three cases: Ne^{20}, Mg^{24}, and Si^{32}. As we shall see in a moment, a large question mark should be put on the state in S^{32}. This question mark has developed within the past year. As yet, the gamma decays from the other states have not been studied, even though in principle they can be. The gamma decay in C^{12} may have been observed, as we shall see in a minute. The nucleus O^{16} is a very interesting case for observing gamma decays since it is a "closed shell" nucleus, so there are no nice M1 levels available for the cascade to go through. Thus, the strength is probably spread around and it may well make the experiment difficult.

A. The 1p Shell

I would now like to go through some of the more recent experimental developments on these levels, starting at low mass number and going up. The first case I would like to look at is that of mass eight. This is one of the most interesting examples that we have because it was a case where there was a theoretical prediction as to what experiment should be done and, in fact, the experiment was done and the state was found. And so, this example could give us great hope that one can perhaps predict the properties of these levels. I will come back to this later in the talk. But the point is that the state in Be^8 had been looked for very carefully in the Li^7+ p reactions by a group at Minnesota.[2] The location of the T = 2 level in Be^8 should be known quite well since the twice removed analog He^8 has been observed and its mass is known fairly well and there are rather reliable mass formulas which allow you to extract the location of the level in Be^8. This was done, the level was looked for, and it was not found with the proton reactions. However, Barker and Kumar,[3] working with the adjacent shell model states, predicted that the proton width might very well be small but that the state might have a large deuteron width. Therefore the resonance was looked for by Black and co-workers[4] at Canberra with deuteron reactions. Figure 2 shows some results of the experiment. In the (d,p) reaction they plot the yield of protons forming the excited state of Li^7 normalized to the yield of protons forming the ground state. They had evidence that there was no resonance in the yield of p_0, in agreement with the fact that the proton-induced reactions did not show the resonance. Well, in the p_1 yield they found a definite anomaly at just about the right place. The anomaly is observed at several angles and in the alpha channel as well. So this resonance seems to be reasonably well established and it comes at the proper place to agree with the mass of He^8.

The next case I would like to consider is mass twelve. Many people over the last few years have tried to locate the T = 2 resonance in C^{12} by means of resonance reactions, i.e. compound nucleus reactions, without much success. The state had been seen in the two-nucleon transfer reaction and so the location of the level was fairly well known. The earliest work was done by Cerny *et al.*[5] More recently the level has been substantiated at Cal Tech with observations on the (p,t) two-nucleon transfer reaction. The excitation energy was found to be (27.595 ± 0.020) MeV.[6] But despite these observations all efforts to locate the level with proton reactions or deuteron reactions had failed. Again, on the basis of the shell model states that should lie nearby, Barker and Kumar[7] made a definite prediction of the levels that could mix with the T = 2 level. They predicted that there should be proton and deuteron widths for the state. At least two laboratories, Canberra and Stanford, independently used the deuteron reaction [8,9] as well as the proton reaction[10] to

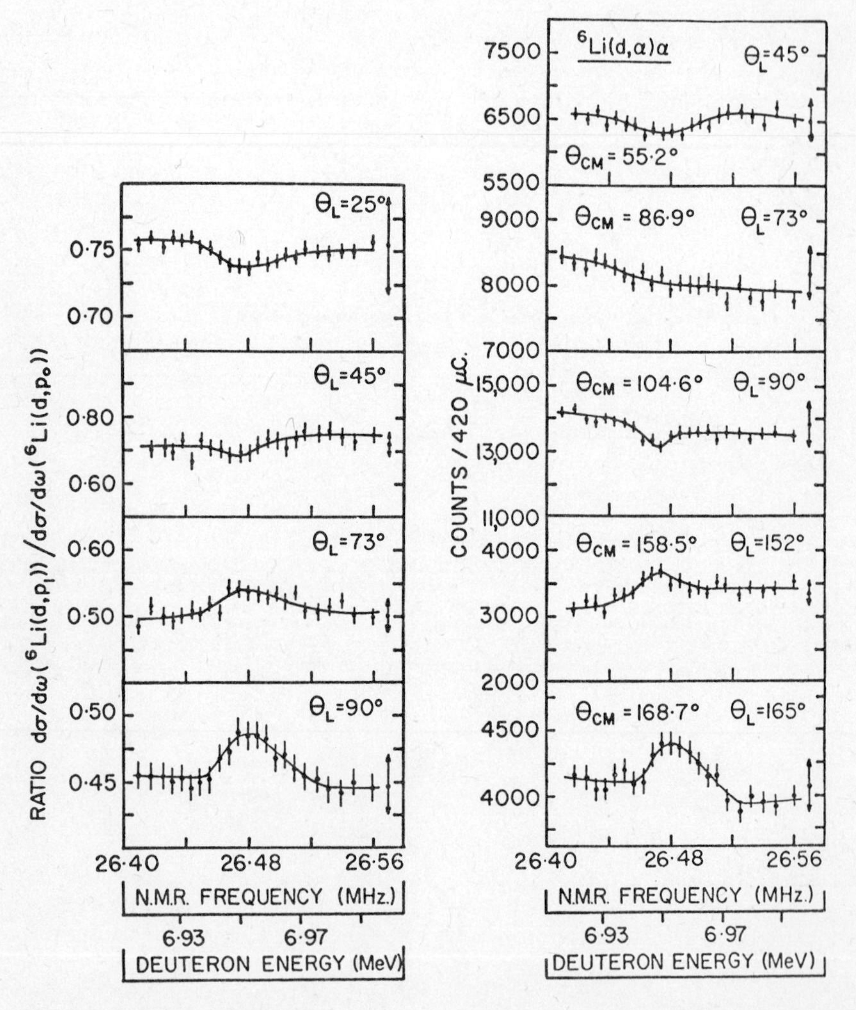

Fig. 2. Left: The $(d,p_1)/(d,p_o)$ ratio in the region of the lowest T = 2 resonance in Be^8. Right: The (d,α) yield normalized to the yield of (d,p_o) at $\Theta_L = 25^o$. The target thickness corresponded to a deuteron energy loss of 8±3 keV. The arrows on the right of each figure show ±5% as a guide to the size of the anomaly in each case. Figure from Ref. 4.

search for this resonance. The first reaction is B^{10}+ d, the second B^{11}+ p. All these investigations proved to be negative. However, very recently, Black and co-workers[8] have found a resonance in the Be^9 + He^3 reaction, which is a new channel that has not received much previous attention in the study of the T = 2 levels. The experiment consisted in observing coincidences in the reaction $Be^9(He^3,\gamma\gamma)$. The T = 2 level in C^{12} almost certainly cascades through the 15 MeV level, the prominent 1^+, T = 1 level in C^{12}, and so you would expect the T = 2 level to be identified by a resonance in the yield of this (γ,γ) cascade. Figure 3 shows that a resonance was observed in three different runs. The authors do not make strong claims as to the identity of this resonance, but it does appear at just the right place to be identified with the T = 2 level, as observed in the two-particle transfer reaction.

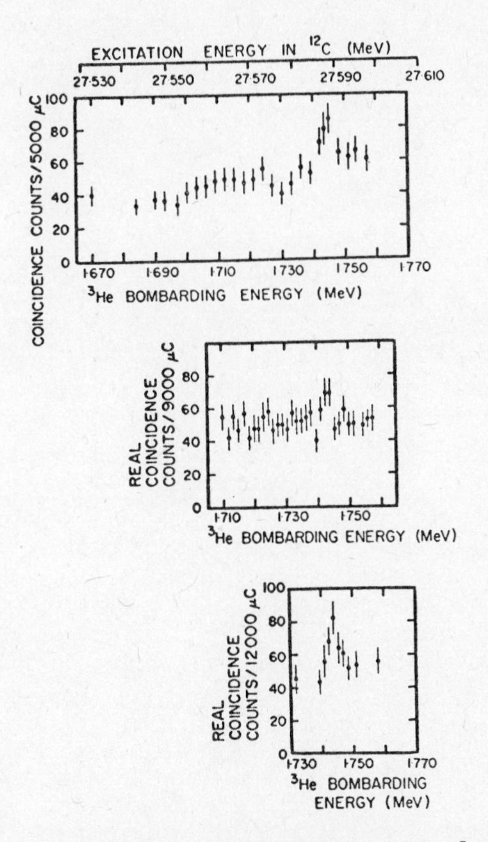

Fig. 3. Three measurements of the excitation function for $Be^9(He^3,\gamma\gamma)$ in the expected excitation energy region for the lowest T = 2 state of C^{12}. The uppermost data were taken with a resolution of 6-8 keV, the lower two sets of data with a resolution of 2-3 keV. Figure from Ref. 8.

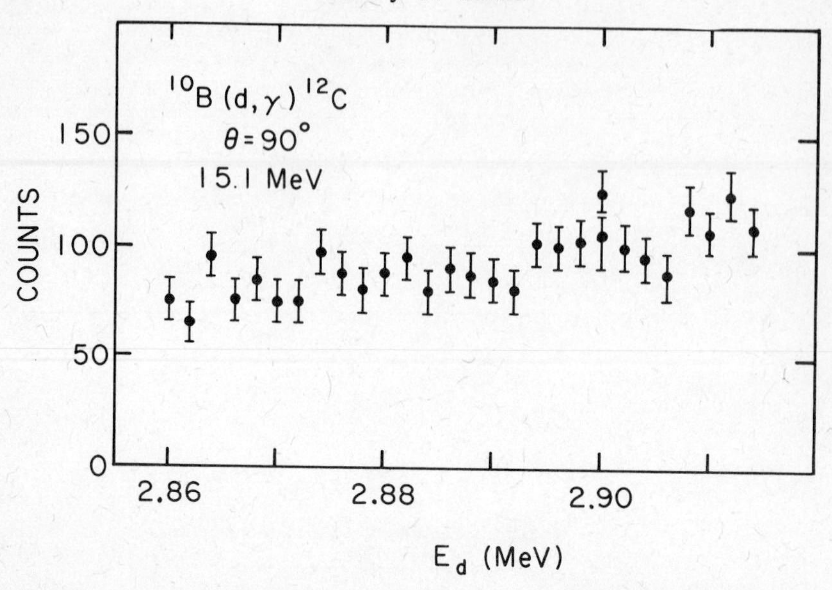

Fig. 4. The yield of 15.1 MeV γ-rays from the cascade in $B^{10}(d,\gamma)C^{12}$ in the region of the expected 0^+, T = 2 resonance of C^{12}. Data from Ref. 9.

Figure 4 shows the search for this resonance made with the $B^{10}(d,\gamma)$ reaction at Stanford by Chang and co-workers.[9] There is an interesting experimental point in this measurement. I think experimentalists will recognize that if there is any B^{11} in the target you will observe many 15-MeV γ rays from the very strong $B^{11}(d,n\gamma)$ reaction. So, in order to look for the resonance by observing 15-MeV γ rays (singles detection) a B^{10} target made at Argonne by the mass separator was used. In such a target the B^{11} content is completely negligible and certainly much less than anything in a separated target of nominal 99.9% purity. As can be seen in Figure 4 there is a yield of 15-MeV γ rays - this yield is not just background, but is due, we believe, to some kind of a direct capture process or to a very broad resonance. The expected location of the T = 2 resonance is at $E_d \simeq 2.88$ MeV. Since we do not know the parameters of this resonance yet, it is difficult to say what limit this experiment places on the deuteron strength, but just from general experience with these reactions, I think one could say that the deuteron width is probably less than 10% of the total radiation width.

We turn next to the mass sixteen case. The T = 2 resonance in O^{16} had been observed at Stanford[11] by a transfer reaction, but there had not been much success in resonance reactions. Recently, the resonance has been observed in both the deuteron reactions and the alpha reactions.[12] Figure 5

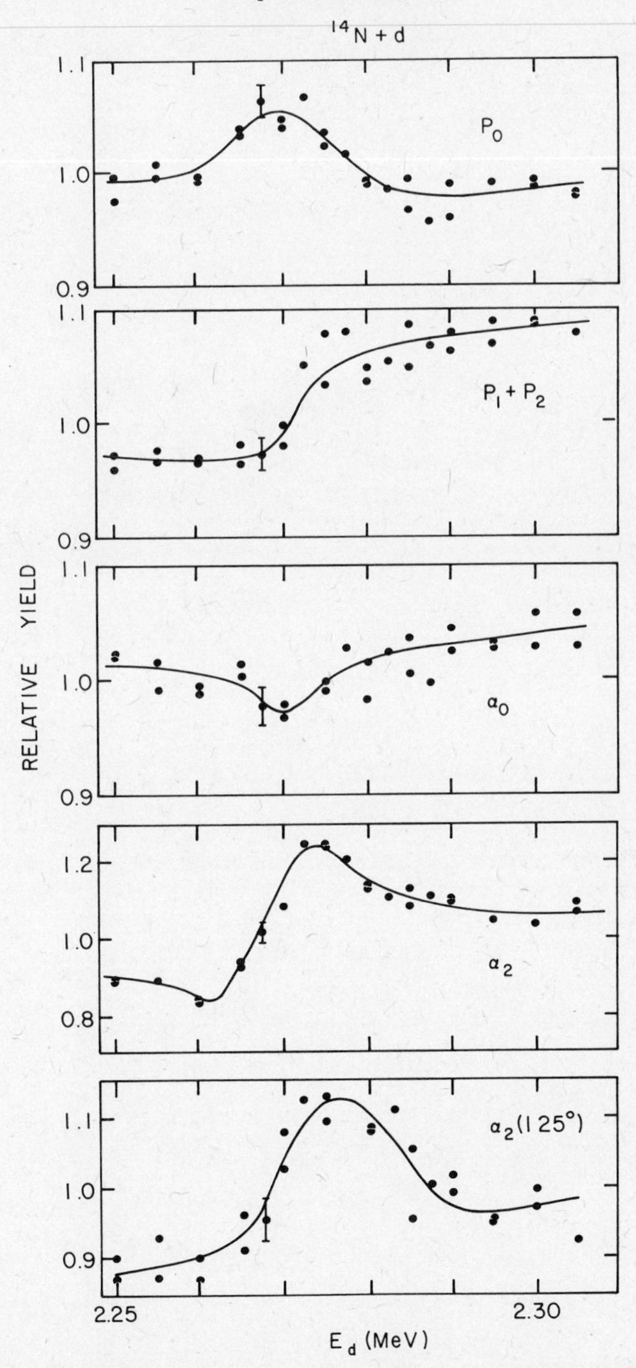

Fig. 5. Yield of particles from N^{14} + d in the region of the expected 0^+, T = 2 resonance of O^{16}. Data from Ref. 12.

shows results of the N^{14} + d reactions. The resonance shows up in several channels and is particularly strong in α_2. We should recall that α_2 corresponds to the 0^+ level of C^{12}, not the ground state but the next 0^+ level. Thus, the T = 2 level appears to have a strong connection with this 0^+ level in C^{12}; in fact, you might want to turn the argument around and say that that is evidence for the fact that the level in O^{16} is the 0^+, T = 2 level.

Figure 6 shows the alpha work. You can see that very strong resonances are obtained in the (α,α_2) reaction (all the curves in the left of the figure). The p_0 channel also shows strength.

So, as we finish the evidence in the 1p shell let us summarize it briefly. There appears to be no nice regularity in the partial widths of these T = 2 levels in the 1p shell. Some of them have deuteron widths and no observable (as yet) proton widths. Others may have proton widths and no deuteron widths. Some have alpha widths and perhaps one of them even has a fairly large He^3 width. Although the statistics are not very great, it is certainly very hard to make any kind of regularity out of this. The information in the 2s-1d shell, which I would now like to sketch briefly, supports this general observation.

B. The 2s-1d Shell

A lot of the experimental evidence in the 2s-1d shell has already been published, so I would like to review only some of the highlights.[13] In touching the highlights I want especially to bring out some of the experimental difficulties and remaining problems. The best studied level[10] is the T = 2 level in Ne^{20}. The level diagram in Figure 7 more or less summarizes what is known about Ne^{20} as far as the high isospin levels are concerned. So far there is only one M1 transition observed connecting the ground level with any higher T = 1 level. The best evidence comes from electron scattering work by Fagg and co-workers at NRL.[14] They find no strength to any other 1^+, T = 1 level and, in a sense, that is supported, although it is not direct support by any means, by the fact that the 0^+, T = 2 level shows only one cascade through the same level. These statements are all meant to apply within experimental observation and error. The nucleus Ne^{20} is also the only case in which the first excited T = 2 level has been observed and studied. This level was found at an excitation energy which agrees very closely with the excitation energy of the first excited state in O^{20}. As in the p shell, both these T = 2 levels seem to be well endowed with particle widths of all kinds. We will now look at several examples.

The neutron channel is open for the second T = 2 level,

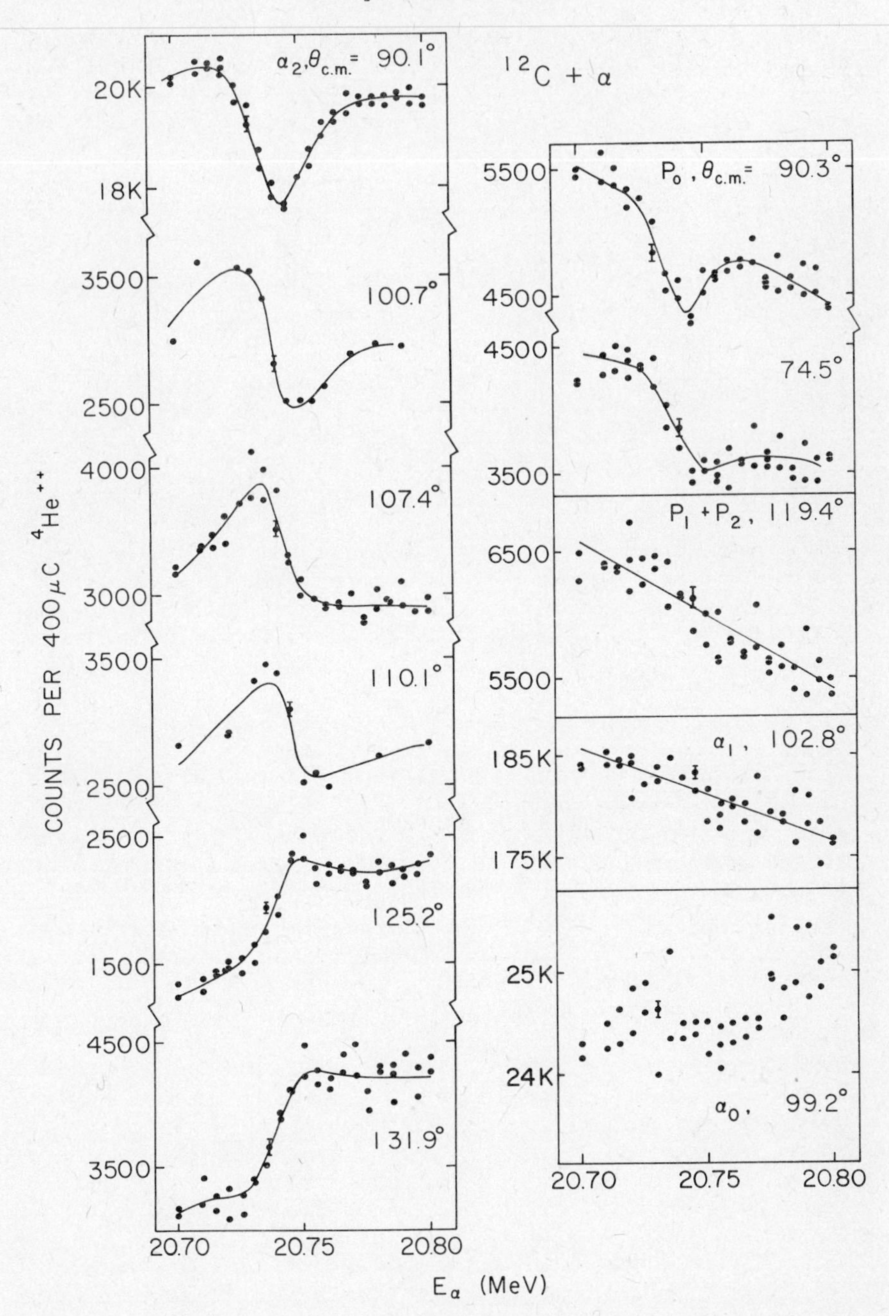

Fig. 6. Yield of particles from C^{12} $+\alpha$ in the region of the expected 0^+, T = 2 resonance of O^{16}. Left: The (α,α_2) yield at several angles. Right: Yield of other particles at various angles. Data from Ref. 12.

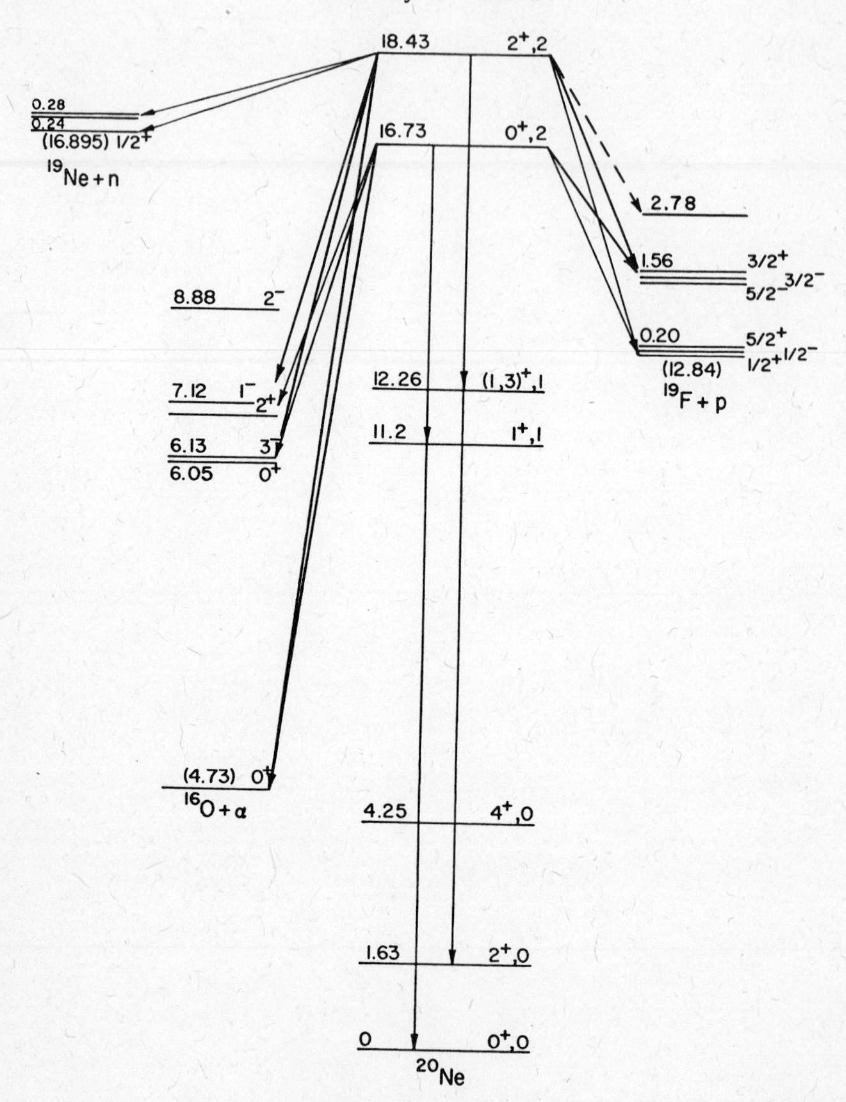

Fig. 7. Decay properties of the two lowest T = 2 levels of Ne^{20}. The assignment of the level at E_x = 12.26 MeV is uncertain.

so we have the chance of measuring not only the proton width but also the neutron width which provides an additional means of observing the operation of the isospin selection rule. The (p,n) resonances have been observed[15] and are shown in Figure 8.

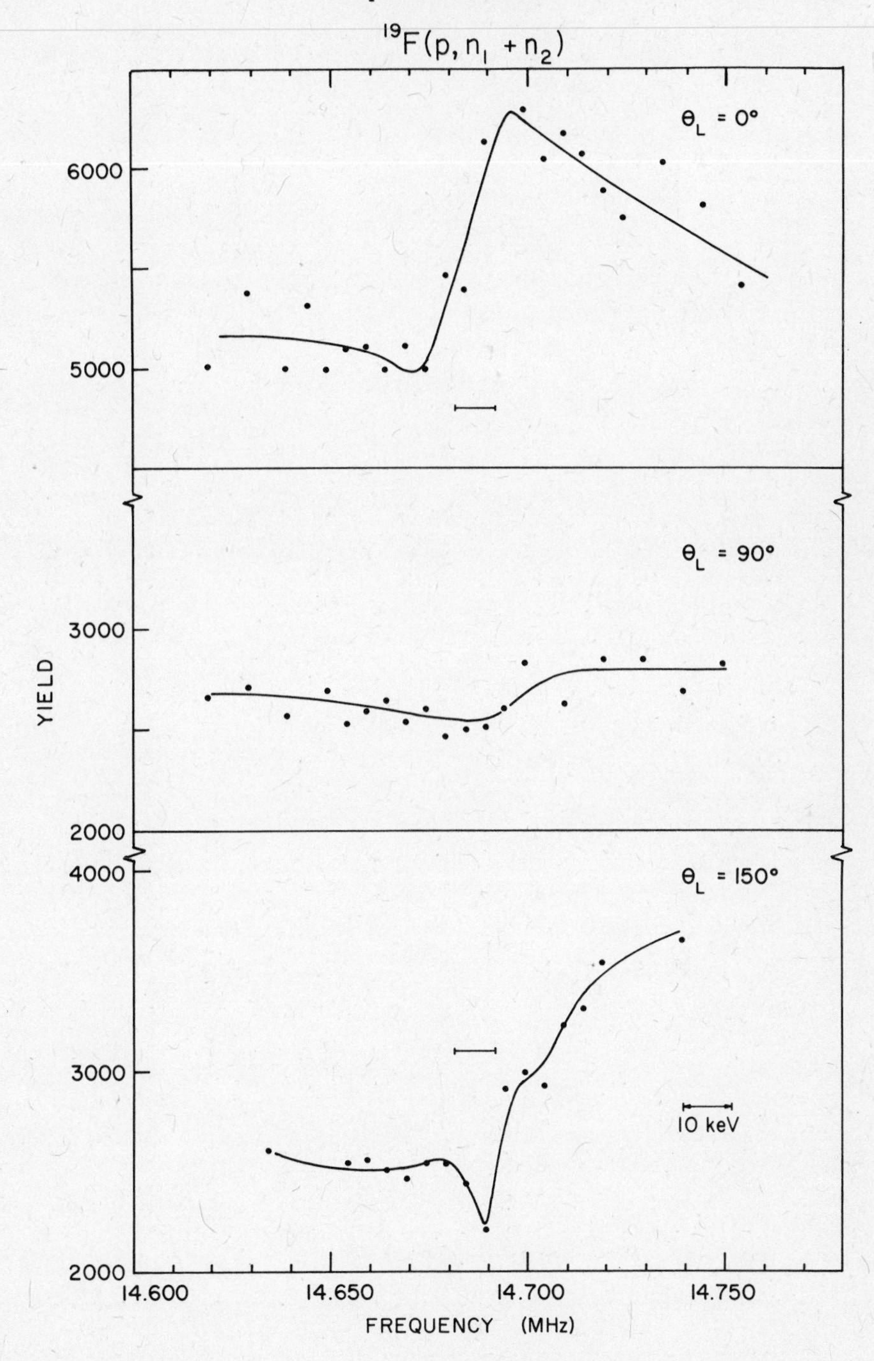

Fig 8. Yield curves from the reaction $F^{19}(p,n_1 + n_2)Ne^{19*}$ at the second T = 2 resonance in Ne^{20}. Figure from Ref. 15.

Figure 9 shows some recent, high resolution work at Stanford[16] on the lowest T = 2 level of Ne^{20}. The proton elastic scattering resonance is shown at the bottom of the figure. The figure is meant to illustrate that there are alpha decay widths to the ground state of O^{16}, the next 0^+ state, the 3^- state, and the 1^- and 2^+ states as well. So even though the spins and parities of these levels range from 0^+ to 3^-, the widths seem to be comparable in all these cases. The analysis of these curves is in progress and hopefully one will be able to obtain values for the partial widths, although one should beware of the inherent dangers and pitfalls in such analyses.

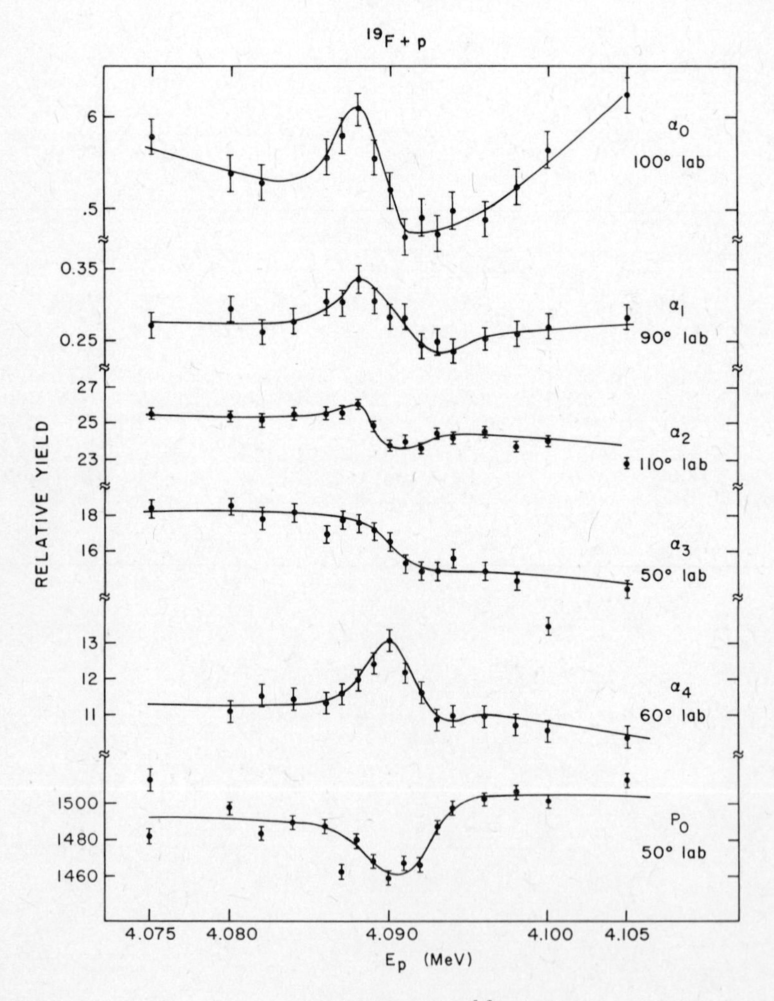

Fig. 9. Yield of particles from F^{19} + p in the region of 0^+, T = 2 level of Ne^{20}. Figure from Ref. 16.

Table I summarizes the data[10] on the T = 2 state in Ne^{20}. I only want to show that there is a fair amount of information on this level. The total width is about 2 keV. The number for $\Gamma_p\Gamma_\gamma/\Gamma$ is about 1 eV, from which one can extract the radiation width Γ_γ if one knows the proton branching ratio, but that is a rather difficult number to get. It is obtained from the analysis of the resonance scattering curve, and the best information that we have on this at the moment is that it is about 0.06 which gives a radiation width for this level of about 15 eV with a rather large error.[17]

Attention should be drawn also to the result

$$\Gamma_{\gamma 1}/\Gamma_\gamma < 0.15$$

listed in this table. This limit is the result[17] of four separate measurements, two with a 7 keV target and two with a 2.5 keV target. The failure to observe the cross-over $0^+, 2 \rightarrow 2^+, 0$ transition is consistent with the fundamental

$\Delta T \neq 2$

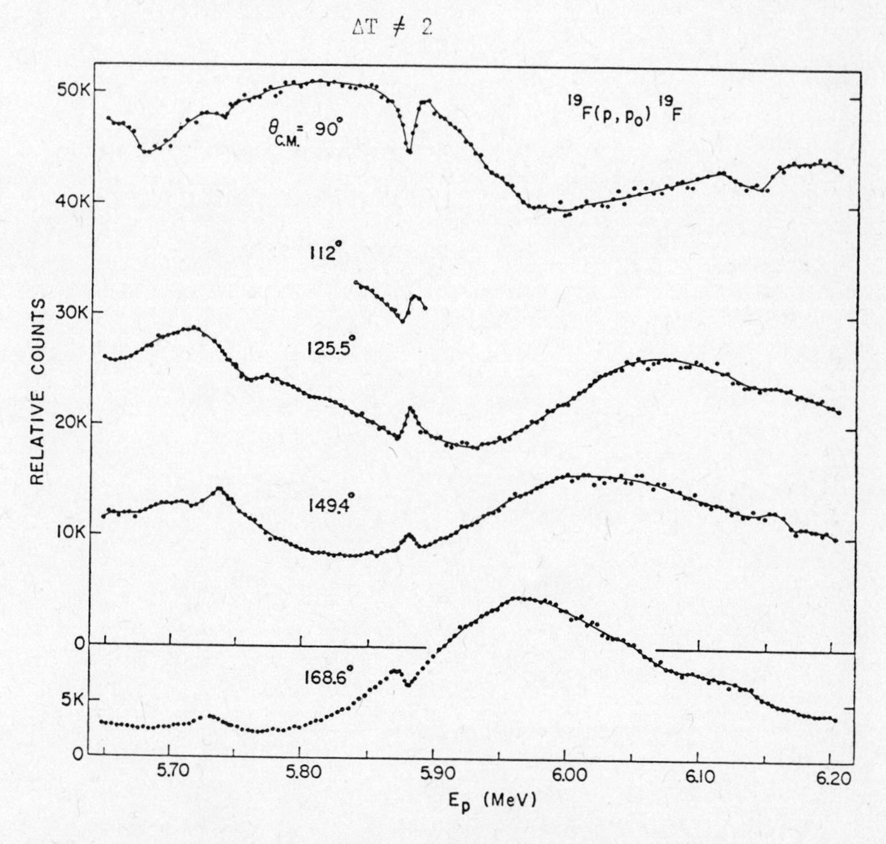

Fig. 10. Elastic scattering of protons at the second T = 2 state in Ne^{20}. Figure from Ref. 18.

Table I

Properties of the Lowest $(0^+,2)$ Level of Ne^{20}

Properties	Reference
$E_x = 16.8 \pm 0.1$ MeV	19
$E_x = 16.730 \pm 0.006$ MeV	20
$E_p = 4.096 \pm 0.003$ MeV $(E_x = 16.734 \pm 0.003$ MeV)	21
$\Gamma = 2.1 \pm 0.5$ keV	
$\Gamma_p/\Gamma = 0.031 \pm 0.002$	
$\Gamma_{(p_o+p_1+p_2)}/\Gamma = 0.16 \pm 0.15$	22
$\Gamma_{(p_3+p_4+p_5)}/\Gamma = 0.14 \pm 0.15$	
$\Gamma_{(\alpha_1+\alpha_2)}/\Gamma = 0.38 \pm 0.15$	
$\Gamma_{(\alpha_3+\alpha_4)}/\Gamma = 0.32 \pm 0.15$	
$E_p = 4.090 \pm 0.005$ MeV $(E_x = 16.729 \pm 0.005$ MeV)	16,23

Table I
(continued)

Properties	Reference
$\Gamma \lesssim 2$ keV	
p_3, p_4, p_5 not observed	
$\alpha_o, \alpha_1, \alpha_2, \alpha_3, \alpha_4$ observed	
$E_p = 4.090 \pm 0.005$ MeV ($E_x = 16.729 \pm 0.005$ MeV)	10
$\Gamma = 2 \pm 1$ keV	
$\Gamma_p \Gamma_\gamma / \Gamma = 0.92 \pm 0.23$ eV	
$\Gamma_\gamma \simeq 15$ eV	
$\Gamma_{\gamma_1} / \Gamma_\gamma < 0.15$	
$\Gamma_{\alpha_2} \simeq \frac{1}{2}(\Gamma_{\alpha 3} + \Gamma_{\alpha 4})$	
$(\Gamma_{\alpha 3} + \Gamma_{\alpha 4})/\Gamma_\gamma \lesssim 100$	

selection rule for γ-radiation discussed below.

In Figure 10 is shown the elastic proton scattering at the second T = 2 resonance[18] in Ne^{20}. The interference patterns in these curves are consistent with d-wave scattering with $\Gamma_p/\Gamma \approx 0.2$ (for $J^\pi = 2^+$). This resonance also shows widths for p_3, p_4, p_5, α_0, $\alpha_1 + \alpha_2$, α_3, and α_4.

As already indicated, for the second excited T = 2 level, the neutron channel is open and it is interesting to look for the neutron decays which can be done with a time-of-flight spectrometer. Figure 8 shows work done at Stanford.[15] In this case, you see that there are nice (p, $n_1 + n_2$) resonances at the second excited T = 2 level. The (p, n_0) resonance is also observed.

Figure 11 shows another example of the operation of the $\Delta T \neq 2$ selection rule, in this case, for transitions from the second excited T = 2 level of Ne^{20}. Here you can look for both cross-over transitions, the one to the ground state which would be E2 and the one to the first excited level which would be M1 + E2. The experiment has not been done very carefully but within statistics the cross-over radiations are not resonant.[24]

Figure 12 shows the situation in Mg^{24} where a T = 2 level was first observed by resonance reactions.[25] I only want to point out two features. One is that for the first time a splitting of the M1 strength is observed. Both for transitions from above and below the M1 strength is split between two levels. Also, I want to point out that in studying these gamma ray transitions one can always be guided experimentally by the analog beta transitions if they are known. For each strong beta transition one expects to find a strong analog M1 gamma transition. In all cases this has been so and, in fact, it then becomes interesting to compare the strength of these analog beta and gamma transitions in order to determine the amount of the orbital contribution to the M1 matrix element. We will return to this a little bit later. We should note in this case there is an observed analog beta decay from the (1^+, 1)-level which is not the ground state. Usually in the 2s-1d shell the ground state of the $T_z = 1$ nucleus is not the (1^+, 1)-level.

I would like to pass on to Si^{28} where the experimental situation becomes more difficult, since the levels are in an excitation region around 15 MeV where the level density is becoming quite high. The level density of $T_<$ levels with low spin is not necessarily a concern because you are still looking for very sharp levels and they stand out if they are located on broad overlapping levels. The difficulty now is that you begin to see very many high-spin states which are narrow because of the angular momentum barrier and they can be very easily confused with the sharp T = 2 levels. In both the Si^{28} and the S^{32} cases the experimental

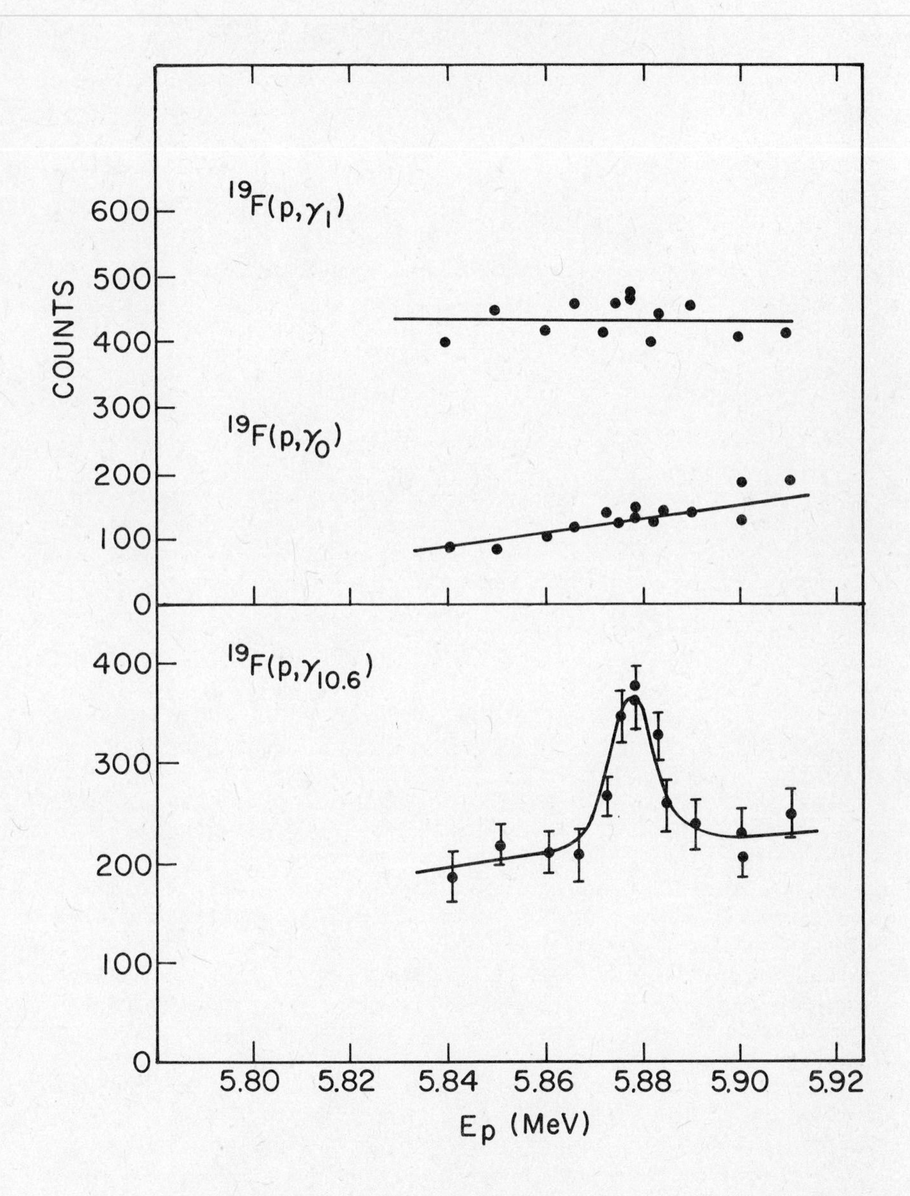

Fig. 11. Yield of γ-rays from F^{19} + p in the region of the 2^+, T = 2 level of Ne^{20}. The second γ-ray in the isospin allowed cascade is designated $\gamma_{10.6}$; the two forbidden ($\Delta T = 2$) cross over γ-rays to the ground and first excited states are labeled γ_0 and γ_1. Figure from Ref. 24.

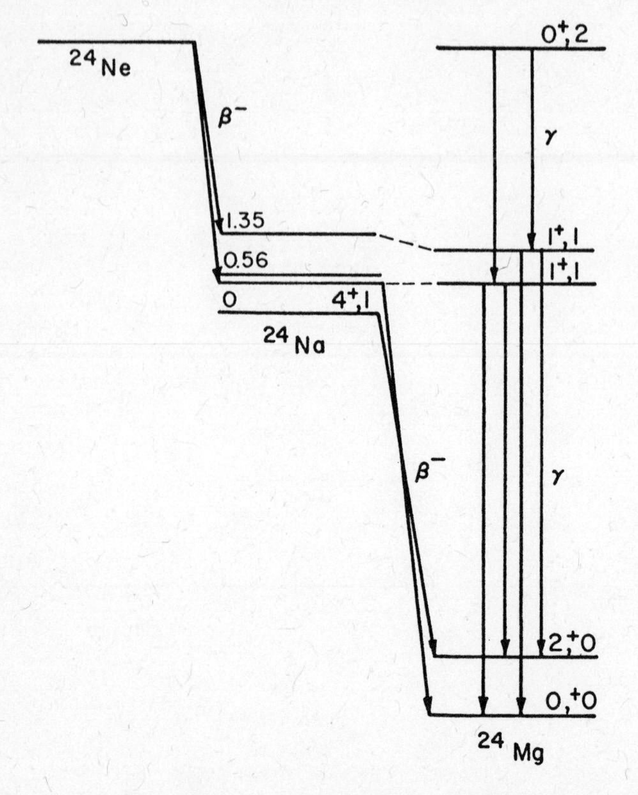

Fig. 12. Level scheme of A = 24 nuclei showing the analog beta (GT) and gamma (M1) transitions.

situation has been rather confused. I think it has been straightened out in Si^{28} but not completely in S^{32}.

In Si^{28}, despite a long search,[26] no resonance in the (p,γ) reaction was observed which qualified for the T = 2 level. However, the level was found to have a large alpha width from observing alpha decays following particle transfer reactions.[22] So, the resonance was looked for and found in (α,γ) capture.[26] The results are shown in Figures 13 and 14.

In Figure 13 the most significant feature displayed is that the sharp (p,γ) resonance in this region does not coincide with the (α,γ) resonance. The former resonance has been shown to have properties in disagreement with a $(0^+,2)$ assignment.[10]

Since the decay scheme shown here was established with a sodium iodide detector one cannot be sure whether there are two or three T = 1 levels taking part in the cascade, but the decay of these levels apparently involves at least two or three gamma rays. This decay scheme has been verified in part by Thwaites *et al.*[27]

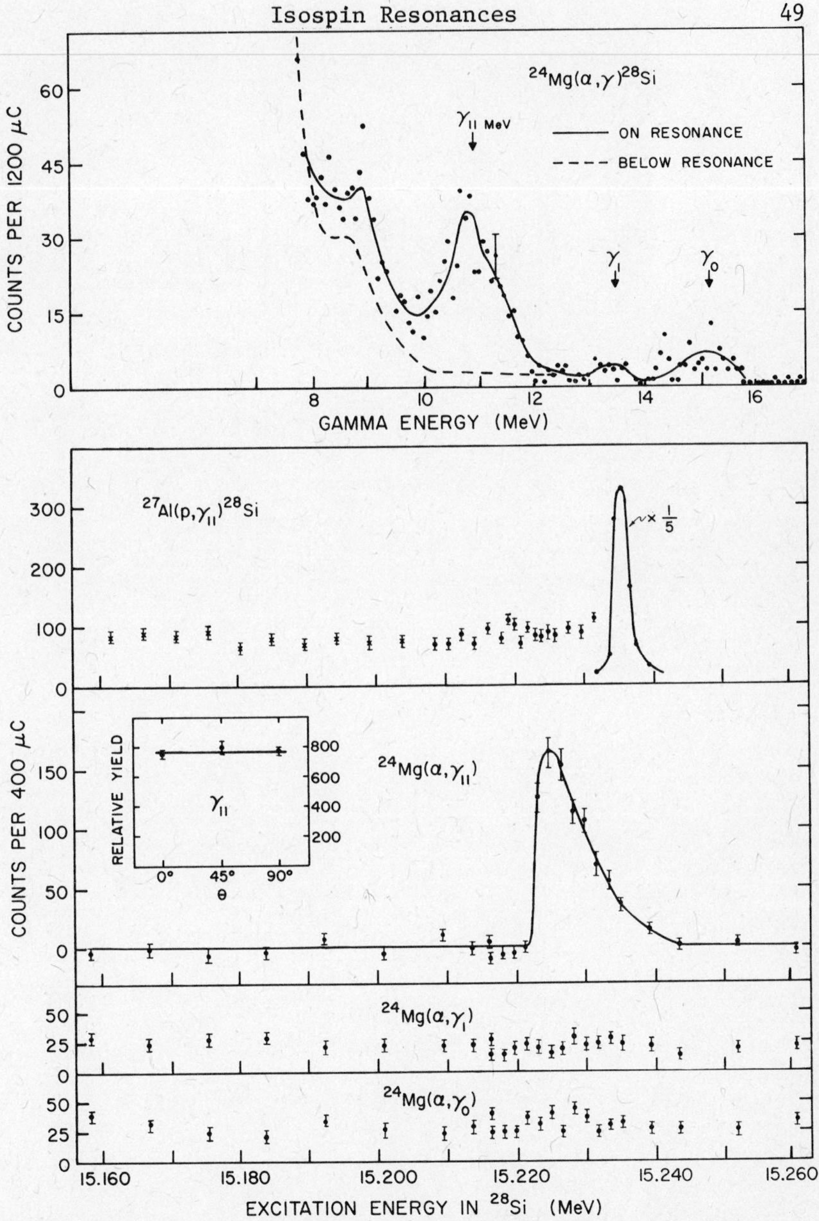

Fig. 13. Spectra and yield curves from $Mg^{24}(\alpha,\gamma)Si^{28}$. Bottom three curves: The yields of γ_0, γ_1, and the 11 MeV gamma-ray group, as a function of excitation energy in Si^{28}, produced by the reaction $Mg^{24}(\alpha,\gamma)Si^{28}$ (10 keV Mg^{24} target). Insert: The on-resonance angular distribution of the 11 MeV group. Top curve: The solid line is the sum of three partial gamma-ray spectra at the resonance in $Mg^{24}(\alpha,\gamma)Si^{28}$; the dashed line is the sum of off-resonance spectra. Second curve from top: The yield of gamma-rays near E_γ = 11 MeV from the reaction $Al^{27}(p,\gamma)Si^{28}$; the dots are for a 2 keV Al^{27} target, and the crosses for a 5 keV target. Figure from Ref. 26.

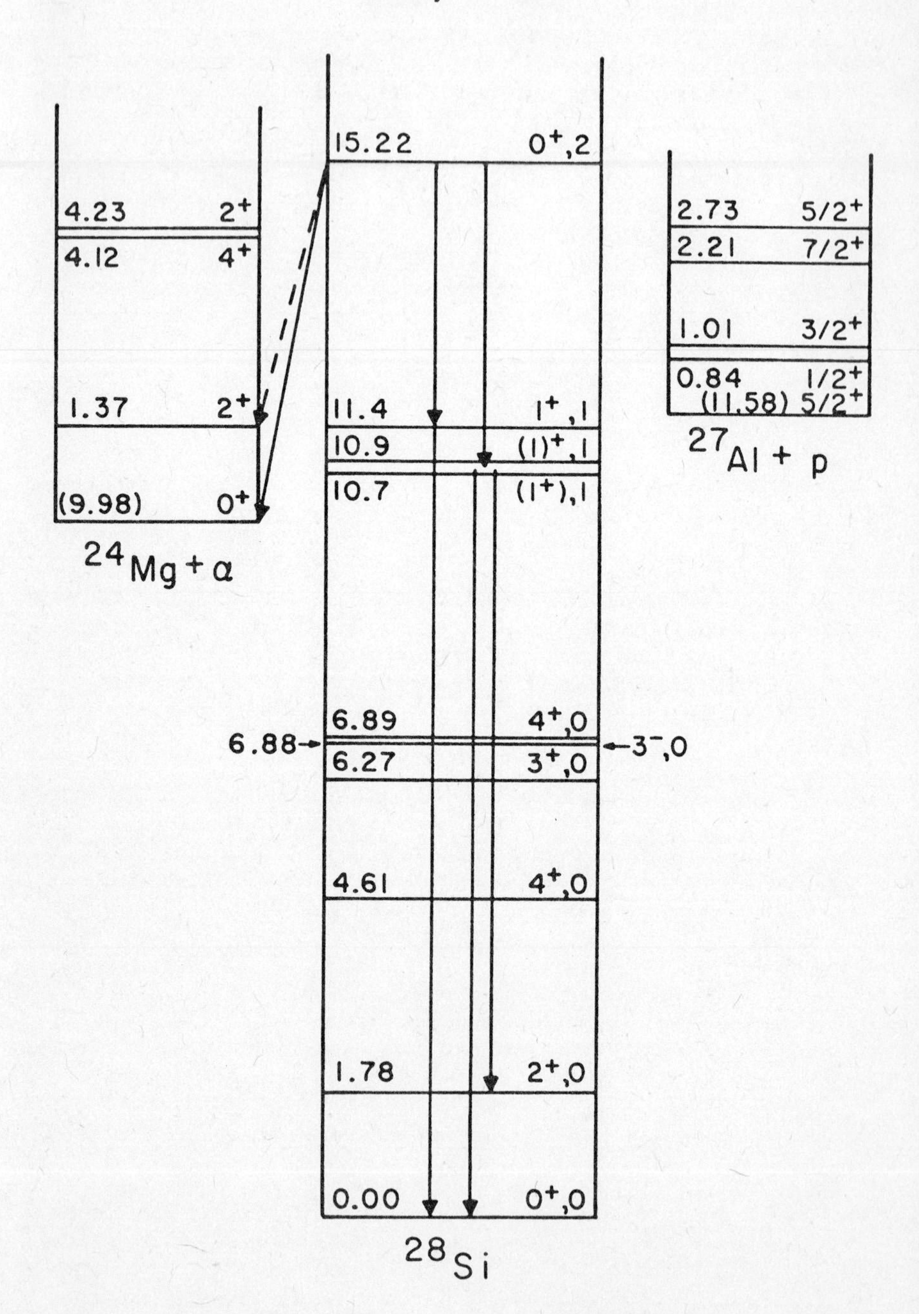

Fig. 14. Level diagram of Si^{28} showing the observed decay modes of the $(0^+, 2)$-level. The level structure at 10.8 MeV is uncertain. Figure from Ref. 10.

It is interesting to note that there is also a sharp resonance nearby, formed in alpha capture, which is observed to decay primarily by cascade through the rotational levels of Si^{28}, probably including the 6^+ level.[27] Therefore, it is interesting to speculate that this level might be a high-spin state connected with the rotational band of the ground state. This resonance can be seen in alpha scattering[10] at E_α = 6.065 MeV in Figure 16.

Figure 15 shows the alpha scattering at the E = 6.115 MeV resonance, which corresponds to the (α,γ) resonance in Figure 13. Resonances in (α,α) are observed at the three angles where the P_2, P_4, and P_6 coefficients should vanish for spins 2, 4, and 6, respectively, in a $(0^+, 0^+)$-scattering process. Thus, we can eliminate these spins, as well as odd spins which should not resonate at $\theta_{lab} = 90^o$. At the lower resonance, E_α = 6.065 MeV, the resonances are observed at these same angles. So one can say that both these resonances have either spin 0^+ or 8^+ or greater. Therefore, since the upper resonance appears to decay to a T = 1 level, we identify it with the 0^+, T = 2 level, and speculate that the other level, which appears to decay through the rotational levels, may indeed be an 8^+ or 10^+ level. We hope that further work will clarify this.

It is presumed that the very strong resonance at E_α = 5.995 (Figure 16) is a 0^+, T = 0 level whose proximity might account for the mixing which produces the α width in the T = 2 state.

Table II summarizes the information on the T = 2 level of Si^{28}.

The situation in S^{32} has been confused, since there appear to be several candidates for the T = 2 level. The Berkeley group[22] observed the lowest T = 2 state in S^{32} as a final state in the isospin allowed reaction $S^{34}(p,t)S^{32}$ at an excitation energy of $E_x = 12.034 \pm 0.040$ MeV. Yield curves observed[10] in the reaction $P^{31}(p,\gamma)S^{32}$ in this region of excitation energy are shown in Figure 17. The 7-MeV gamma ray could arise either from transitions to the 3^- state at 5.01 MeV in S^{32}, or from the second member of a cascade in isospin T = 2 → T = 1 → T = 0. The 7.7-MeV gamma ray could come either from transitions to the 4.29-MeV 2^+ state, or from an isospin cascade through a possible T = 1 level at around 7.7 MeV.

At the lower resonance (E_p = 3.178 MeV) the presence of a strongly resonant γ_1 is not consistent with a T = 2 assignment. At the upper resonance (E_p = 3.281 MeV), coincidence measurements proved that the resonant 7-MeV gamma ray is the population gamma ray for the 5.01-MeV, 3^- state and hence the level is not a T = 2 level. At the middle resonance (E_p = 3.221 MeV), coincidence measurements indicated a 5 MeV-5 MeV-2 MeV triple gamma ray cascade,

Table II

Properties of the Lowest $(0^+,2)$ Level of Si^{28}

Properties		Reference
$E_x = 15.206 \pm 0.025$ MeV		22
$\Gamma_{\alpha_o}/\Gamma = 0.81 \pm 0.10$		
$\Gamma_{\alpha_1}/\Gamma = 0.09 \pm 0.04$		
$\Gamma_{\alpha_0}/\Gamma = 0.05 \pm 0.09$		
$\Gamma_{p_1}/\Gamma = 0.05 \pm 0.06$		
$E_\alpha = 6.118 \pm 0.003$ MeV	$(E_x = 15.224 \pm 0.005$ MeV$)$	27
$E_\alpha = 6.115 \pm 0.003$ MeV	$(E_x = 15.221 \pm 0.005$ MeV$)$	10

Table II
(continued)

Properties	Reference
$\Gamma \lesssim 2$ keV	
$\Gamma_{\alpha}\Gamma_{\gamma}/\Gamma \simeq 1.25^{+0.25}_{-0.45}$ eV	
$\Gamma_{\gamma} \simeq 1.75^{+0.04}_{-0.7}$ eV	
$\Gamma_{\gamma_1}/\Gamma_{\gamma} < 0.03$	
$\Gamma_{p_o}/\Gamma_{\alpha_o} < 0.08$	
α_o, α_1 observed	
p_o, p_1 not observed	

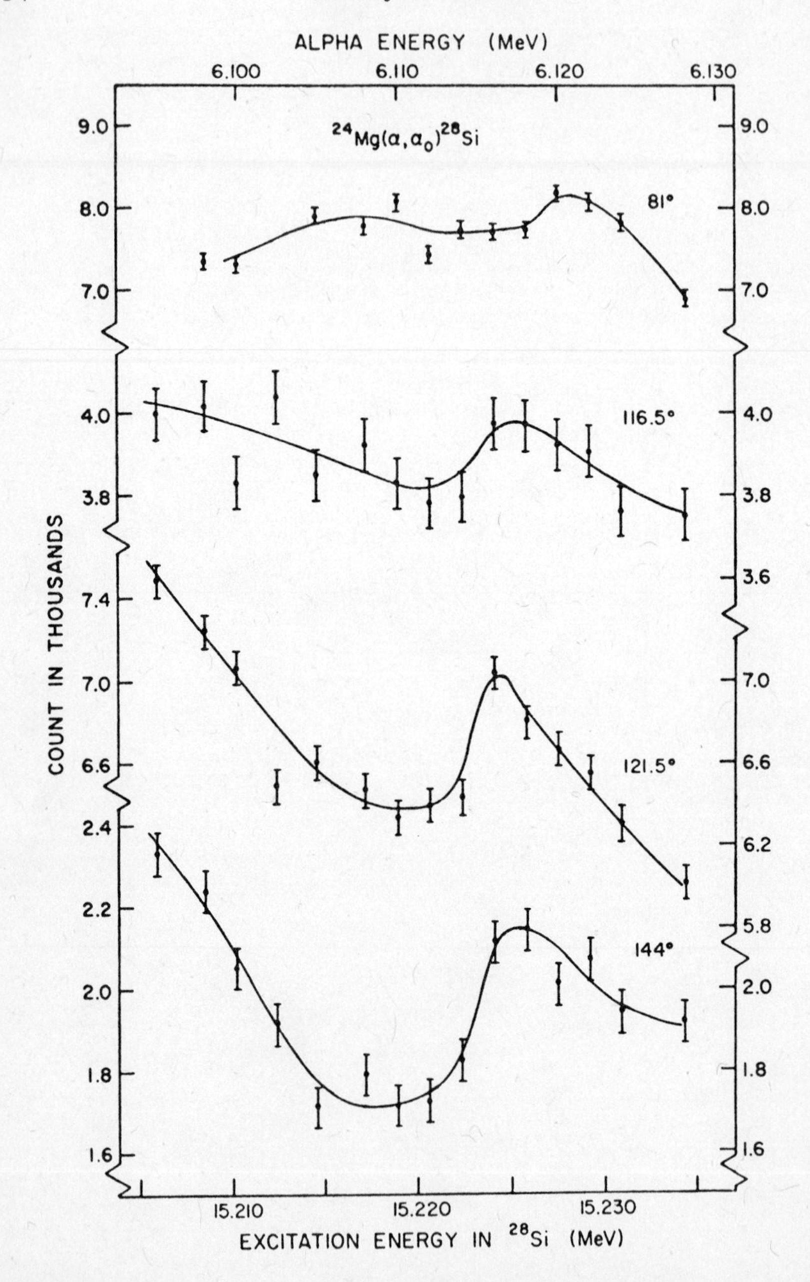

Fig. 15. Elastic scattering yield from $Mg^{24}(\alpha,\alpha)Mg^{24}$ near E_α = 6.115 MeV. The yield curves were measured with a Mg^{24} target 5 keV thick, evaporated onto a thin carbon backing. Figure from Ref. 26.

which is indicative of a T = 2 → T = 1 → T = 0 cascade to the first excited state at 2.23 MeV. However, it has been shown by a group at Orsay[28] that γ-rays emitted following particle decay of this level are not isotropic. This feature very probably eliminates a 0^+ assignment.

The Orsay group has studied the weak resonance at E_p = 3.190 MeV, just above the strong resonance at E_p = 3.178 MeV in Figure 17. Although it has several attractive features, they have finally rejected this level as a candidate for a 0^+, T = 2 on the basis of its decay scheme.[29] They have found another weak resonance about 6 keV above the strong resonance at E_p = 2.381 MeV (Figure 17) which may prove to be the correct resonance.[29]

The yield curves for particle reactions throughout the region of S^{32} discussed above are shown in Figure 18.

I think S^{32} illustrates the great difficulty and amount

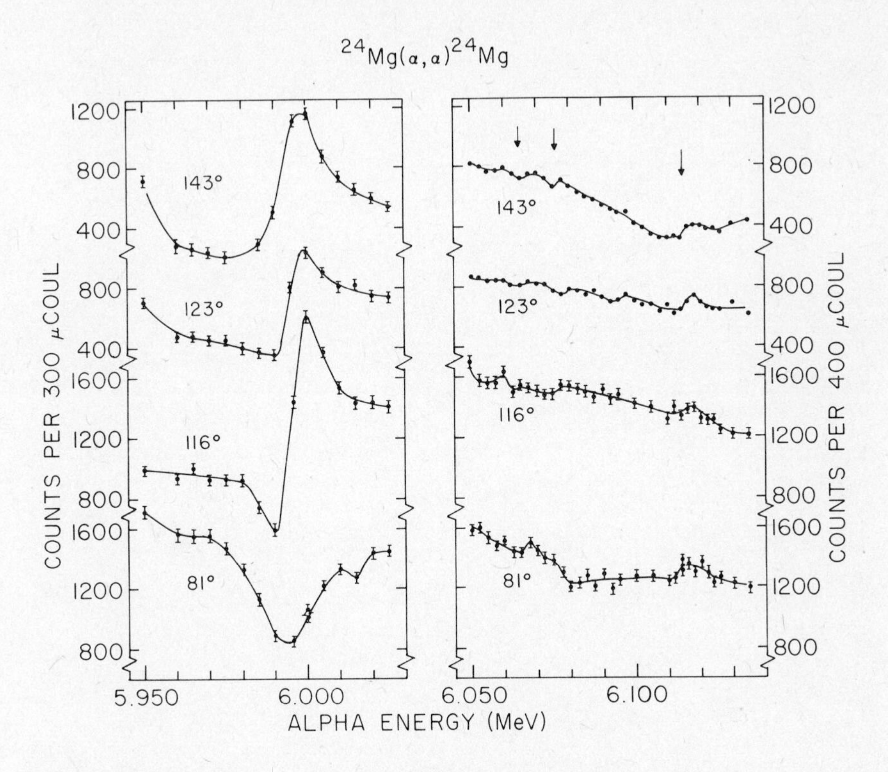

Fig. 16. Elastic scattering yield from $Mg^{24}(\alpha,\alpha)Mg^{24}$ showing resonances in the vicinity of the 0^+, T = 2 resonance at E_α = 6.115 MeV. Attention should be drawn to the strong resonance at E_α = 5.995 MeV and the weak resonance at 6.065 MeV discussed in the text. Figure from Ref. 10.

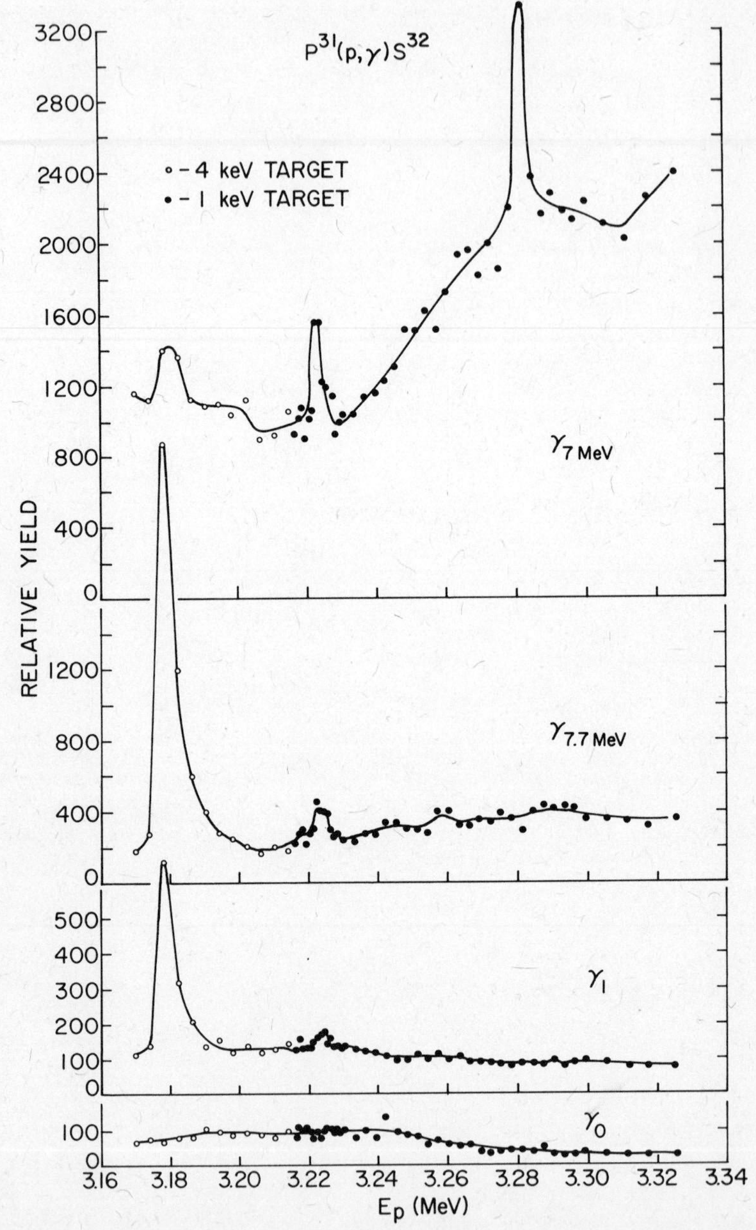

Fig. 17. Yield curves from $P^{31}(p,\gamma)S^{32}$ showing yields of the γ_0,γ_1, 7.7 and 7.0 MeV gamma-rays. The open circles correspond to data measured with a 4 keV target, and the solid circles correspond to data measured with a 1 keV target. Part of the 7 MeV gamma-ray nonresonant yield is due to contaminants. Figure from Ref. 10.

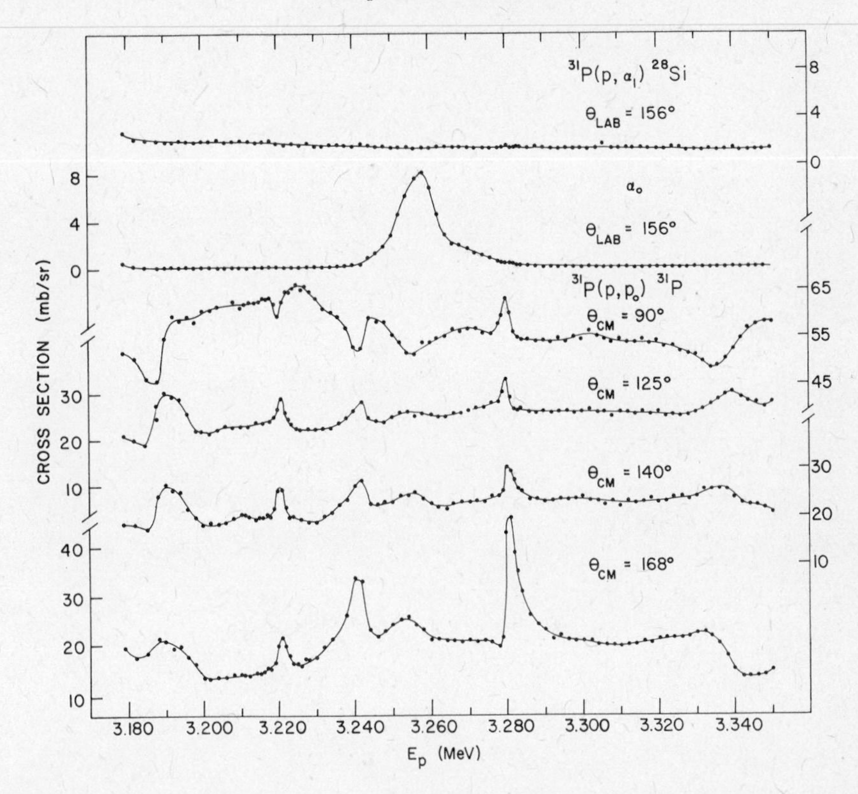

Fig. 18. Yield curves for $P^{31}(p,p_o)P^{31}$, $P^{31}(p,\alpha_o)Si^{28}$, and $P^{31}(p,\alpha_1)Si^{28}$ measured with a one keV target. Figure from Ref. 10.

of work it requires to identify these high isospin resonances in medium weight nuclei where the level density of high-spin levels is high. Nevertheless, detailed properties of the levels can be obtained once they are located in the resonance reactions.

III. THE T = 1 RESONANCES

I would now like to look at the T = 1 levels which are involved in the cascades from the T = 2 levels. Most of the information on these 1^+, T = 1 levels comes from electron scattering, and most of the recent work has come from the group at NRL. I will limit the discussion to the levels in the s,d shell, since the levels in Be^8 and C^{12} have been known for a long time.

The electron scattering from Ne^{20} shown in Figure 19 is very impressive.[14] It shows nothing but the ground state and a big peak at 11.2 MeV and then nothing. Thus the M1 strength is concentrated into a single giant M1 transition, much as it is in C^{12} and Be^8.

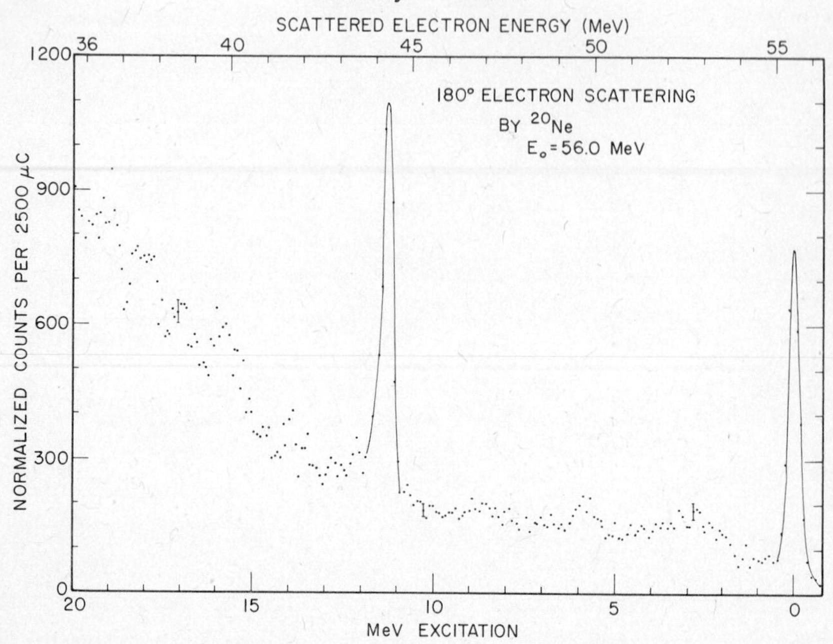

Fig. 19. Inelastic electron scattering from Ne^{20} at 180° and E_o = 56.0 MeV. The M1 strength is concentrated chiefly in the $(1^+, 1)$-state at 11.2 MeV. Data from Ref. 14.

The electron scattering from Mg^{24} in Figure 20 shows the very striking splitting of the M1 strength,[30,31] which the gamma ray cascades also show, and again to rather high accuracy no other comparable strength. To show that this lack of further structure is really significant the electron scattering[32] from Mg^{26} is presented in Figure 21. When you turn off the selection rule in self-conjugate nuclei which inhibits the $\Delta T = 0$ transitions, you see that the M1 strength is spread over many resonances. Presumably a few of these may still be identified with strong $\Delta T = 1$ transitions.

Figure 22 shows the inelastic electron scattering[33] from Si^{28}. The strength now appears to be split into perhaps three or four M1 levels, although the 11.4-MeV level is much the strongest level. Unfortunately, the location of the $(1^+, 1)$-levels in Si^{28} in this region has not been well established from other studies. Peter Endt has informed me that more is known now on Al^{28} and so insofar as you can make analog comparisons between Al^{28} and Si^{28} more information is to study the gamma decays from the T = 2 level with a germanium detector. In any case, the actual structure of the M1 levels in this region would not upset too much the extraction of the total M1 strength.

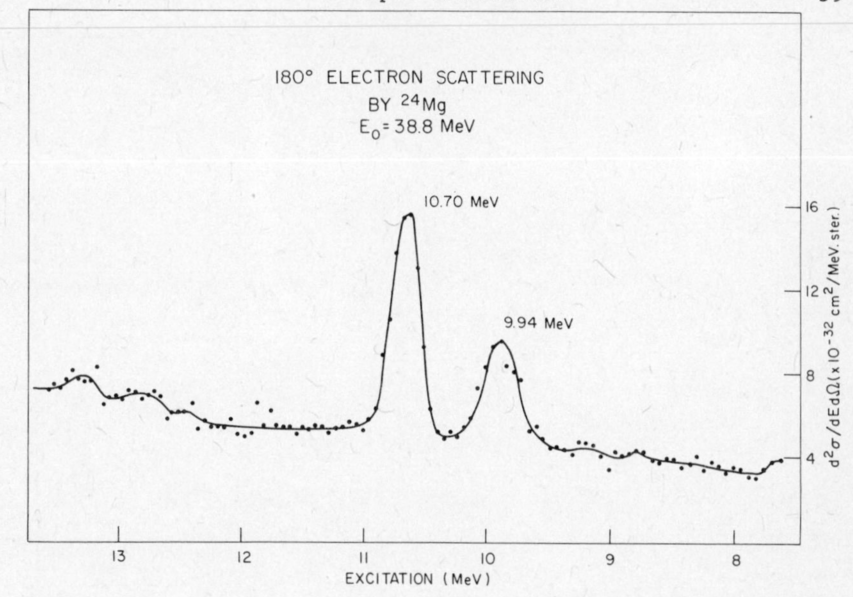

Fig. 20. Inelastic electron scattering from Mg^{24} at 180° and E_o = 38.8 MeV. The M1 strength is concentrated chiefly in the two $(1^+, 1)$-levels at 10.0 and 10.7 MeV. Figure from Ref. 31.

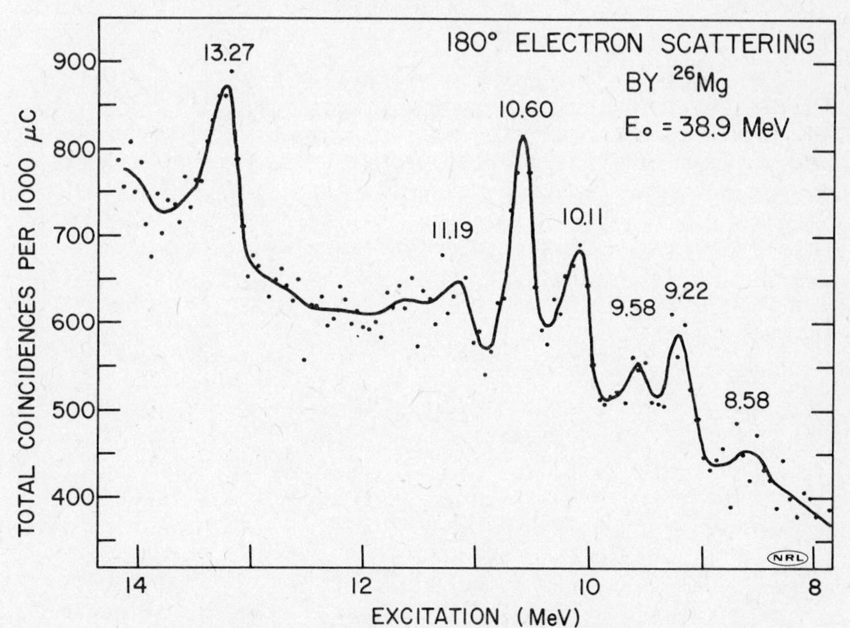

Fig. 21. Inelastic electron scattering from Mg^{26} at 180° and E_o = 38.9 MeV. The M1 strength is spread over many levels in the region from 9-14 MeV. Figure from Ref. 32.

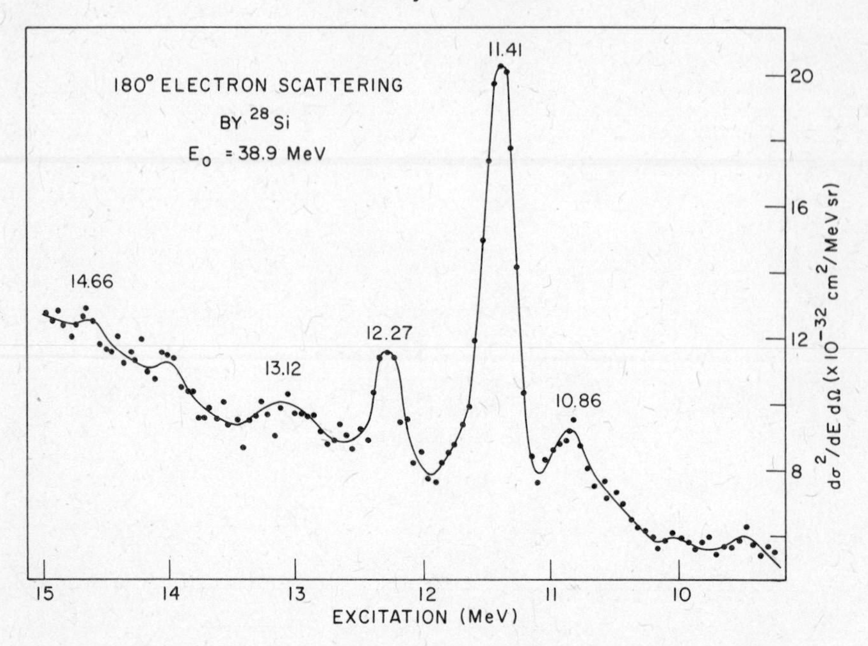

Fig. 22. Inelastic electron scattering from Si^{28} at 180° and E_o = 38.9 MeV. The M1 strength is concentrated in at least three (1^+, 1)-levels at 10.9, 11.4 and 12.3 MeV. Figure from Ref. 33.

A. Application of Sum Rule

Before leaving the T = 1 levels I would like to consider their strengths and see what that tells us about nuclear structure. To do this one can make use of the sum rule given by Dieter Kurath,[34] which says that the energy weighted sum of the M1 strength is given by the expectation value of the $\vec{\ell}\cdot\vec{s}$ term summed over the particles in the ground state. Thus, if you measure the energy weighted sum you then have a measure of $\vec{\ell}\cdot\vec{s}$ in the ground state, and it is very interesting to compute these in the 2s-1d shell where we now appear to have some very good measurements.

In Mg^{24}, if you take the new numbers for the electron scattering,[31] the expectation value for $\vec{\ell}\cdot\vec{s}$ in the ground state comes out to be about 6.8. If Mg^{24} was just a spherical $d_{5/2}$ particles and each contributes one unit (ℓ/2) to the expectation value of $\vec{\ell}\cdot\vec{s}$. The Nilsson model for η = + 4 gives 6.0 so, for what it is worth, the experimental value of the $\vec{\ell}\cdot\vec{s}$ term favors a slightly deformed picture for Mg^{24}.

In Si^{28} it is interesting to look at this sum rule because, as pointed out by Kurath,[34] you might be able to decide whether Si^{28} is oblate or prolate, at least as far as the M1 transitions are concerned. Again, the experimental values are beginning to look more and more reliable, so if you add up the M1 strength in Figure 22, you get a value of 9→10 for the energy weighted sum, depending somewhat on the number of levels included. For $\eta = \pm 4$, the oblate value for the expectation value of $\vec{\ell}\cdot\vec{s}$ is 7.6 and the prolate value is 4.3. Thus the experiment favors either a moderate oblate deformation or a very small prolate deformation (since the spherical value would be 12).

IV. DISCUSSION

Now I would like to go back and pick up some of the loose ends and discuss a few interesting points that we haven't mentioned.

A. Isospin Mixing in Be^8

First of all, I would like to expand on the calculation of Barker and Kumar,[3] because I think it may be the only case so far in which there has been any real success in obtaining the observed isospin mixing. They considered all the 0^+, T = 0 and T = 1 levels in the ($1_s 4$, $1_p 4$)-configuration of mass 8 and obtained five 0^+, T = 0 levels: the ground state, and levels at 19.1 MeV, 28.1 MeV, 29.9 MeV and 40.1 MeV. There are two 0^+, T = 1 at 19.7 MeV and 36.1 MeV, and the 0^+, T = 2 level was calculated at 27.1 MeV. The Coulomb matrix elements which would account for the mixing in the T = 2 level are respectively 33, -14, 51, -6, and 3 for the T = 0 states and 4 and -13 for the T = 1 states. It is natural to attribute the chief mixing to the nearest state, the T = 0 level at 28.1 MeV, which also has the largest matrix element 51. Barker and Kumar then looked at the property of this level and found in fact that this level should not decay into Li^7 + p or Be^7 + n, nor into two alpha particles, nor into Li^5 + t or He^5 + He^3, but should have a very large overlap with deuterons and Li^6. However, despite the apparent success of this calculation, a similar shell model prediction[8] seems to have failed in C^{12}. Thus, although I think this calculation is encouraging and people ought to do more like it, one obviously may not have such a nice simple picture as in Be^8, as we see in the next section.

B. Isospin Mixing in Ca^{40}

The other serious calculation that I know of for T = 2 levels is the one by McGrath et al.[35] They chose Ca^{40}, which is not one of the cases I have presented here, but it is a case known from two particle transfer reactions, and has the very interesting feature that the alpha decay seems to be almost 100%. The other decay channels are less than 3%. These authors tried to explain the alpha width in a rather general calculation in which they mixed in the 0-particle, 0-hole ground state, which is T = 0, the 4-particle, 4-hole excited 0^+ state, T = 0, and then also the 2-particle, 2-hole anti-analog states, which are expected to produce large mixing. We have already remarked that the T = 2 level is basically a 2-particle, 2-hole state in which the configurations are coupled to maximum isospin, T = 2. If you couple these same configurations to T = 1 and T = 0, you have the anti-analog states, and although these states have not been observed in Ca^{40} you can in fact from neighboring nuclei make fairly good predictions about the splitting. McGrath et al. use these predictions and states in a calculation similar to the one I have described above and obtain alpha particle widths, but they are not large enough and, in addition, of course, not correct in detail.

C. Isospin Mixing in Other T = 2 Levels

Rather than give a table of numerical values of all the various partial widths of the T = 2 levels, insofar as they are known, I would prefer just to indicate whether the various widths are strong or weak (or unobserved) as is done in Table III. Although the data are not extensive enough to draw definite conclusions, it would appear that the observed widths occur randomly with about equal probabilities and strengths. Thus there does not seem to be any nice pattern which would indicate a simple origin for these mixings with states of lower isospin.

The alpha widths indicate a mixing of T = 0 levels into the T = 2 states, while the proton widths, of course, require only T = 1 admixtures. At first sight the prevalence of alpha decays, which require an isotensor part in the Coulomb interaction, might seem strange. However, Hecht[36] has pointed out that $\Delta T = 2$ admixtures should in general be expected to be of the same order of magnitude as $\Delta T = 1$ admixtures, i.e., although the isovector interaction is much larger than the isotensor interaction in the diagonal elements of the Coulomb interaction, they are expected to be of the same order of magnitude in the off-diagonal elements.

Hecht and Janecke[37] have studied isospin admixtures in several 1p and 2s-1d shell nuclei. One possibility is to

Table III

Widths of T = 2 levels. The symbol X indicates a width is observed, W that it is weak or unobserved, and - that it is energetically forbidden or has not been looked for.

	p_o	n_o	d_o	α_o	He^3_o
^{8}Be	W	-	X	X	-
^{12}C	W	-	W	-	X(?)
^{16}O	X	-	X	X	-
^{20}Ne	X	-	-	X	-
$^{20}Ne^*$	X	X	-	X	-
^{24}Mg	X	-	-	W	-
^{28}Si	W	-	-	X	-
^{32}S	X	-	-	W	-
^{40}Ca	W	-	-	X	-

invoke the anti-analog states, as done above, to provide the mixing. [In this connection one may recall the celebrated case of the mixing of (2^+, 1) and (2^+, 0) levels in Be^8.] However, this picture would be expected to provide some regularity in the observed widths, instead of the rather random pattern observed experimentally.

At present, the mixing seems to be provided by the chance juxtaposition of a (0^+, 1) or (0^+, 0)-state. Thus, the strong (0^+, 0)-level, observed to lie very close to the (0^+, 2) level of Mg^{24}, (see Figure 16) may provide the mixing in this case.

Many people have suggested[38] that any attempt to obtain the observed mixing by considering only this internal kind of state mixing may be unsuccessful, because the mixing may in fact take place in the external wave functions just outside the nucleus. I think some efforts have been made to explain the mixing on this basis.

D. Shell Model Description of T = 2 Levels

The (1^+, 1)-levels discussed in this talk have been studied rather extensively by means of electron and gamma ray excitation. Transitions to these levels from the (0^+, 0)-ground state are strong M1 transitions and exhaust a large portion of the M1 strength. Thus, the transitions from the (0^+, 2)-level to the (1^+, 1)-levels, which are of comparable strength, might be similar in nature and complete the picture, i.e. the excitation of the (0^+, 2)-level from the ground state is accomplished by means of two strong single-particle M1 jumps.

Kurath[39] has carried out a shell model calculation for Ne^{20} in the complete s-d shell with Pandya's interaction. In this case the level spacings are satisfactorily obtained, and the calculations predict a dominant cascade through the lowest (1^+, 1)-level. The transition strengths in Ne^{20} are given in Table IV. It is seen that the E^3 factor is decisive in picking out the transition to the lowest (1^+, 1)-level from the (0^+, 2)-level, since major reduced strength resides also in two higher levels. On the other hand, in the transitions from the ground level to the (1^+, 1)-levels, the energy factor enhances two higher levels until they are comparable to the lowest (1^+, 1)-levels. These levels have not yet been observed in electron scattering.[14]

The shell model picture remains essentially the same for Mg^{24} and thus is unable to account for the splitting of the M1 strength between the two low-lying (1^+, 1)-levels (in the transition both from above and below). However, Akiyama *et al.*[40] in their shell model treatment do find some justification for the splitting in Mg^{24}.

On the other hand, this feature may be better explained in the deformed model which is discussed next.

Table IV

Calculated M1 gamma strengths in Ne^{20}. The $(0^+,2)$ level is calculated to lie at 16.6 MeV (Ref. 39).

E(11) (MeV)	$\Lambda(02\to11)$	$\Gamma(02\to11)$ (eV)	$\Lambda(00\to11)$	$\Gamma(11\to00)$ (eV)
10.3	17.7	11.8	6.7	6.8
12.2	3.5	0.8	0.04	0.07
13.8	6.7	0.4	0.45	1.1
14.1	2.8	0.1	0.10	0.25
14.3	5.1	0.1	2.05	5.5
16.0	20.3	0.01	0.96	3.6
16.5	38.8	----	0.01	0.05
17.2	5.1	----	0.02	0.09

E. Rotational Model

The Nilsson level diagram for Ne^{20} is shown in Figure 23. All possible M1 transitions from the ground state are indicated by arrows. Kurath[39] has calculated the strengths of these transitions. The transitions indicated by dashed arrows are expected to be very weak as they vanish in the asymptotic limit. The other two transitions are chiefly orbital in nature. For a value of $\eta = +4$ Kurath obtains a strength

$$\Lambda(M1) = 13.4$$

for the transition from level 6 to level 7, and

$$\Lambda(M1) = 0.6$$

for the transition from level 6 to level 9. Thus, the observation of a single strong transition to the lowest $(1^+, 1)$-level, as observed in electron scattering, is confirmed. To complete the picture, if the $(0^+, 2)$ analog level of Ne^{20} is described by elevating two particles to the 7 orbit, then for $\eta = +4$ the transition $(6^2 7^2)$ to $(6^3 7)$ is the only one possible and has a strength

$$\Lambda(M1) = 8.9.$$

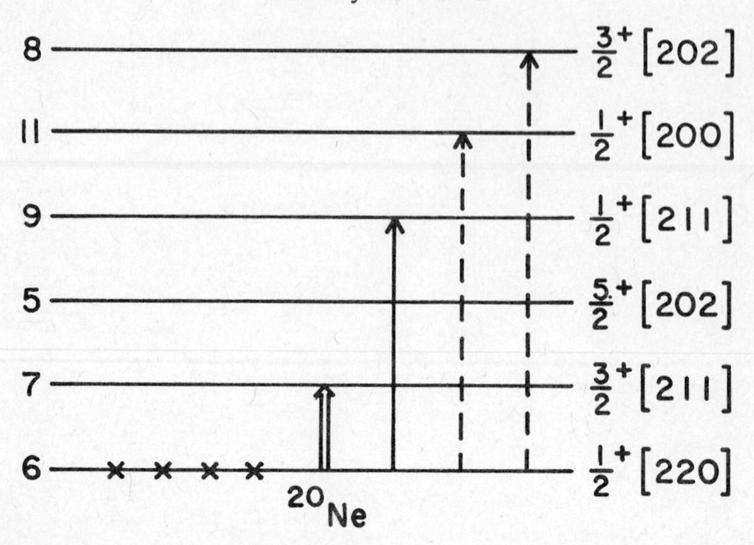

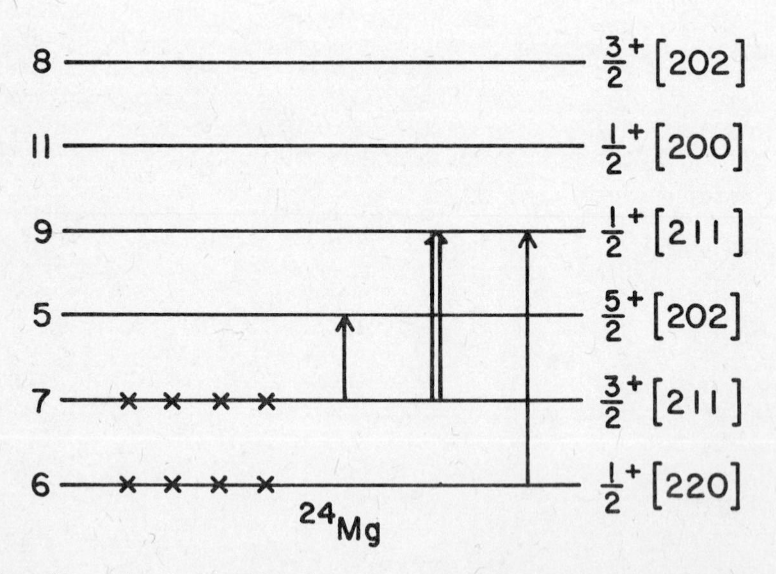

Fig. 23. Top: The M1, $\Delta T = 1$ transitions in the deformed model of Ne^{20} (Ref. 38). Bottom: The M1, $\Delta T = 1$ transitions in the deformed model of Mg^{24} (Refs. 30 and 38).

The Nilsson level diagram for Mg^{24} is given in Figure 23. In this case we have the two orbital M1 transitions $(7\rightarrow5)$ and $(6\rightarrow9)$ and also the spin flip transition $(7\rightarrow9)$. As pointed out by Kuehne et al.,[30] the spin flip transition, effective for both protons and neutrons, should be the strongest transition and may be identified with the transition from the ground state to the second $(1^+, 1)$-level. The transition to the lowest $(1^+, 1)$-level is then identified as $(7\rightarrow5)$. These assignments reproduce the relative strengths of the two transitions, as observed in gamma and electron excitation, but the calculated absolute strengths are too big. Nevertheless, the basic picture appears to be correct. In this picture, if the $(0^+, 2)$-level is chiefly $(7^2 5^2)$, then the transition $(7^2 5^2)\rightarrow(7^3 5)$ will dominate, as observed experimentally in the decay of the $(0^+, 2)$-level.

F. Search for $\Delta T = 2$ Gamma Transitions

In the observed alpha decay of the $T = 2$ levels, the total isospin changes by two units. It is also of considerable interest to determine if a $\Delta T = 2$ gamma transition can be observed. The result of such a search would test the basic rule that $\Delta T = 0, \pm 1$ for all gamma transitions.

One way to break this rule would be to have some kind of exchange current in nuclear forces involving a pion with $T = 1$, so that if the radiation flips the isospin you can have a $\Delta T = 2$ transition. On the other hand a $\Delta T = 2$ transition would rule out the quark picture of nucleons and pions, since the quarks only have isospin 0 or ½. These arguments have been explicitly stated by de Shalit.[41]

A favorable nucleus in which to search for a $\Delta T = 2$ gamma transition at present appears to be Mg^{24} for which the alpha widths of the $T = 2$ level are considerably smaller than in Ne^{20}, for example. The search involves looking for a resonance in the $Na^{23}(p,\gamma)$ yield which would correspond to an E2 transition to the $(2^+, 0)$-first excited state since the $(0^+, 2) \rightarrow (0^+, 0)$ ground state transition would be forbidden by spin conservation. In the region of the $T = 2$ resonance the background transition γ_1 is relatively weaker in Mg^{24} than in Ne^{20}, which again favors the selection of Mg^{24}.

The search for this cross-over transition has been made[10] and it is found that

$$\Gamma(0^+ \rightarrow 2^+) < 0.04 \text{ eV}.$$

The Weisskopf estimate for this E2 transition is 1.9 eV. Thus, the $\Delta T = 2$ transition is less than 2% of a Weisskopf unit.

A similar search has been carried out in Si^{28} with a

comparable negative result.

The searches made for the two T = 2 states in Ne^{20} have already been discussed.

G. Analog Beta and Gamma Transitions

I now turn to the question of the beta analogs. Figure 12 illustrates a typical case in the 2s-1d shell where one can study beta analogs from T = 2 states one nucleus removed from the gamma analogs. In the following formula, given by Kurath,[13] I would like to show the simple relationship between the M1 reduced matrix element and the Gamow-Teller reduced matrix element:

$$\Lambda(M1) = 11 \frac{CG_{\gamma}^{2}}{CG_{\beta}^{2}} \left[1 + 0.11 \frac{\langle \ell \rangle}{\langle s \rangle} \right]^{2} \Lambda(GT) \qquad (1)$$

The CG are known Clebsch-Gordon coefficients, and $\langle \ell \rangle$ and $\langle s \rangle$ are the orbital and spin parts of the matrix element, respectively. So if you can measure $\Lambda(M1)$ and $\Lambda(GT)$ you have a measure of the quantity in the brackets, that is, you can determine the importance of the ℓ part of the M1 transition relative to the spin part. If, in fact, the spin part predominates then you have, in the proper units, simply that $\Lambda(M1)$ is equal to $\Lambda(GT)$. The ratio of these quantities is shown in Figure 24. The data are now becoming plentiful enough to make such a figure useful. If you ignore the points for mass nine you can say that elsewhere in the p shell, mass six, twelve, and thirteen, the indication is that the transition is largely spin and there is little orbit involved. In mass nine the beta transitions have become weak which already says the spin part is weak, but the M1 transition does, in fact, turn out to be strong because the ℓ part is still strong. In the 2s-1d shell the majority of cases indicate that the ℓ part of the matrix element plays a very prominent role, or to put it another way, the spin part is weak since the *ft* values tend to get larger than in the p shell.

Comparison of these beta and gamma analogs has been an active field in recent years and it is hoped that future work will shed more light on the subject.

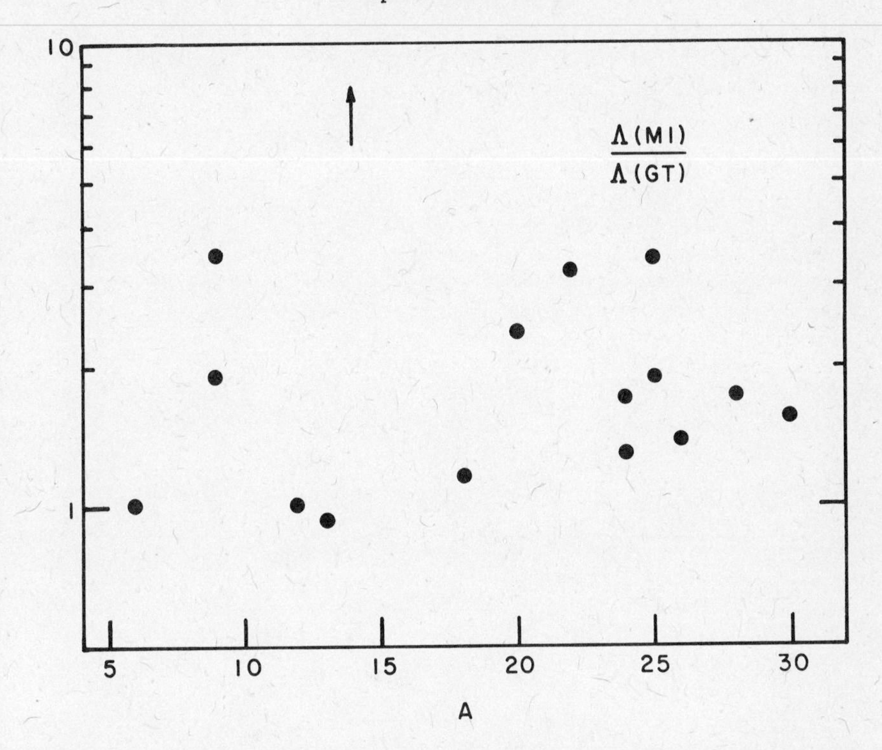

Fig. 24. Comparison of gamma and beta analogues in allowed decays. The quantity in brackets in Eq. (1) is plotted for each A. Experimental values of Λ(M1) and Λ(GT) are used.

V. ACKNOWLEDGMENTS

I am deeply indebted to my colleagues at Stanford, past and present, who have carried out many of the investigations reported in this talk. These colleagues include E. G. Adelberger, C. C. Chang, D. W. Heikkinen, H. M. Kuan, A. V. Nero, W. J. O'Connell, R. E. Pixley, F. Riess, K. A. Snover and B. A. Watson. The pursuit of the T = 2 levels, interesting in itself, has been made even more enjoyable by my association with these colleagues.

REFERENCES

1. S. S. Hanna, in *The Structure of Low-Medium Mass Nuclei*, ed. J. P. Davidson (University Press of Kansas, Lawrence, Kansas, 1968) p. 88.

2. W. D. Harrison, A. R. Barnett, C. Bergmann, and D. Weisser, Bull. Am. Phys. Soc. 13, 1387 (1968).

3. F. C. Barker and N. Kumar, Phys. Letters 30B, 103 (1969).

4. J. L. Black, W. J. Caelli, D. L. Livesey, and R. B. Watson, Phys. Letters 30B, 100 (1969).

5. J. Cerny, Ann. Rev. Nucl. Sci. 18, 27 (1968).

6. C. A. Barnes, D. C. Hensley, P. H. Nettles, and C. D. Goodman, 1970, private communication.

7. F. C. Barker and N. Kumar, quoted in Ref. 8.

8. J. L. Black, W. J. Caelli and R. B. Watson, Phys. Rev. Letters 25, 877 (1970).

9. C. C. Chang, A. V. Nero, C. Kirkham and S. S. Hanna, 1970, private communication.

10. K. A. Snover, thesis, Stanford University, 1969, unpublished.

11. E. G. Adelberger, A. V. Nero, and A. B. McDonald, Nucl. Phys. A143, 97 (1970).

12. A. V. Nero, R. E. Pixley, and E. G. Adelberger, Bull. Am. Phys. Soc., Division of Nuclear Physics, Houston, Texas (1970).

13. S. S. Hanna, in *Isospin in Nuclear Physics*, ed. D. H. Wilkinson (North-Holland, Amsterdam, The Netherlands, 1969) Ch. 12.

14. W. L. Bendel, L. W. Fagg, E. C. Jones, Jr., H. F. Kaiser, and S. Numrich, Bull. Am. Phys. Soc. 13, 1373 (1968).

15. E. G. Adelberger and A. V. Nero, Progress Report, Nuclear Physics Laboratory, Stanford University, 1968, unpublished.

16. B. A. Watson, D. W. Heikkinen, K. A. Snover, S. L. Tabor, and S. S. Hanna, Progress Report, Nuclear Physics Laboratory, Stanford University, 1969, unpublished.

17. K. A. Snover, 1970, private communication.

18. H. M. Kuan and D. W. Heikkinen, Progress Report, Nuclear Physics Laboratory, Stanford University, 1968, unpublished.

19. G. T. Garvey, J. Cerny and R. H. Pehl, Phys. Rev. Letters 12, 726 (1964).

20. E. G. Adelberger and A. B. McDonald, Phys. Letters 24B, 270 (1967) and *Erratum*, Phys. Letters 24B, 618 (1967).

21. R. Bloch, R. E. Pixley, and P. Truol, Phys. Letters 25B, 215 (1967).

22. R. L. McGrath, Second Conf. on Nuclear Isospin, Asilomar, California, 1969.

23. H. M. Kuan, D. W. Heikkinen, K. A. Snover, F. Reiss, and S. S. Hanna, Phys. Letters 25B, 217 (1967).

24. W. J. O'Connell and G. L. Latshaw, Progress Report, Nuclear Physics Laboratory, Stanford University, 1968, unpublished.

25. F. Riess, W. J. O'Connell, D. W. Heikkinen, H. M. Kuan, and S. S. Hanna, Phys. Rev. Letters 25B, 217 (1967).

26. K. A. Snover, D. W. Heikkinen, F. Riess, H. M. Kuan and S. S. Hanna, Phys. Rev. Letters 22, 239 (1969).

27. T. T. Thwaites, P. Kupferman and S. Slack, Bull. Am. Phys. Soc. 14, 566 (1969).

28. S. Gales, M. Langevin, J. M. Maison and J. Vernotte, Institut de Physique Nucleaire, Orsay, 1969, preprint.

29. J. Vernotte, 1970, private communication.

30. H. W. Kuehne, P. Axel, and D. C. Sutton, Phys. Rev. 163, 1278 (1967).

31. L. W. Fagg, W. L. Bendel, R. A. Tobin and H. F. Kaiser, Phys. Rev. 171, 1250 (1968). L. W. Fagg, W. L. Bendel, S. K. Numrich, and B. T. Chertok, Phys. Rev. C1, 1137 (1970).

32. W. L. Bendel, L. W. Fagg, R. A. Tobin and H. F. Kaiser, Phys. Rev. 173, 1103 (1968).

33. L. W. Fagg, W. L. Bendel, E. C. Jones, Jr., and S. Numrich, Phys. Rev. 187, 1378 (1969).

34. D. Kurath, Phys. Rev. 130, 1525 (1963).

35. R. L. McGrath, J. Cerny, J. C. Hardy, G. Goth, and A. Arima, Phys. Rev. C1, 184 (1970).

36. K. T. Hecht, 1967, private communication.

37. See J. Janecke, in *Isospin in Nuclear Physics*, ed. D. H. Wilkinson (North-Holland, Amsterdam, The Netherlands, 1969) Ch. 8.

38. A. K. Kerman, 1969, private communication.

39. D. Kurath, 1969, private communication.

40. Y. Akiyama, A. Arima and T. Sebe, Nucl. Phys. A138, 273 (1969).

41. A. de Shalit, Summary Talk, Heavy Ion Conference, Heidelberg, 1969.

DISCUSSION

WILDENTHAL (Chairman): All right, I think we have time for a few questions before lunch. While people formulate their own, I have one. This problem in S^{32} seems to hinge upon identifying which of the 7-MeV levels is J = 2 and which is J = 1. I think that the $P^{31}(d,p)P^{32}$ reaction of the analog of these two states has been looked at and I think that an MIT-Scandanavia combine have looked at (He^3,d) to the states in S^{32}, and I would guess that perhaps by looking at the (d,p) and the (He^3,d) results to the two sets, one could establish which one of those two 7-MeV states is J = 2 and which is J = 1.

HANNA: That's a nice suggestion. You think they had enough resolution? (Wildenthal answered, "Yes, I think so".) I'm pretty sure the MIT people did, they have a magnet.

WILDENTHAL: The older (d,p) data I am not so sure of, but it strikes me that they did make an assignment as to which one was J = 1 and which one was J = 2.

SANTO: I think they cannot determine the ℓ-values unambiguously, since the angular distributions are rather uncharacteristic at these excitation energies.

KUAN: I just wondered about the cascades from the two possible candidates of the T = 2 state of S^{32}. You mentioned that from the cascades of the candidate at lower energy it is difficult to assign T = 2 to it. How about the candidate at higher energy? Has anybody measured its cascades, using germanium [detectors]?

HANNA: Yes, at Orsay they looked at both the levels I discussed with germanium and they find the decay scheme that I showed. One of the levels decays to the lower T = 1 level and the higher resonance goes to the higher T = 1 level.

KUAN: Do you have a similar difficulty assigning T = 2 to this higher resonance?

HANNA: Yes, the gamma rays following particle decay are not isotropic and also there is a weak transition to the T = 0 first excited state.

TITTERTON: The Canberra $Be^9(He^3,\gamma\gamma)$data locates a C^{12} resonance and 5 keV at 27.585 MeV. We cannot be sure that this is the T = 2 state in the absence of resonances in the B^{10}(d,p), (d,d) and (d,$\gamma\gamma$) reactions. It could be, however, that the T = 2 state has a substantial width for decay into channels so far not studied--such as deuterons to excited states of B^{10} or α-particles to higher excited states of Be^8. Is this possibility being followed at Stanford?

HANNA: Not yet because we haven't been able to make the resonance in the first place. To look at the excited channels you must have one open channel to make the resonance. The He^3 channel may be the only open channel. But, as you say, you can't yet be sure that that makes the right resonance.

ZAMICK: In O^{16} does anybody know anything about T = 1 positive parity states--[do] you know where the T = 0 and T = 2 states are?

HANNA: That was part of my remark about O^{16}. Experimentally, I don't think there are any well established 1+, T = 1 levels that I know of.

ZAMICK: Not necessarily 1+, but say 0+, two-particle, two hole.

HANNA: No, I don't think so.

HARVEY: In connection with the energy weighted sum rule for M1 transitions can I ask whether this is appropriate for both the spin and orbital part of the M1 operator? Would this energy weighted sum of M1 transitions actually give zero if the g.s. has the LS-coupled structure for an even-even nucleus?

KURATH: It's the isovector part, that would be the spin part except you can't calculate the orbital contribution in that case.

ENDT: Would there be ever any hope to see the T = 2 states decaying to 1- states? Or would they always be too high such that they just don't come in?

HANNA: I don't quite know what to predict, but certainly O^{16} would be a good place to look for such transitions since there are good 1- levels there.

WILDENTHAL: Doesn't one have problems with the particle decay channels in O^{16}?

HANNA: Yes, you certainly have the same trouble that you have in Be^8 that the T = 1 levels are unbound to protons and neutrons. These isospin allowed decays make it difficult. However, you should be able to look for the primary gamma ray.

II.A. REACTION MATRIX THEORY FOR NUCLEAR STRUCTURE*

T.T.S. Kuo
State University of New York
at Stony Brook

I. INTRODUCTION

Since the early work of Brown and Kuo[1-4], there has been much interest in using a realistic nucleon-nucleon interaction in nuclear structure calculations. Much work has been done in this area, and the result is generally very promising in the sense that a large number of the detailed properties of nuclei can indeed be explained quite accurately from first principles. In all these calculations, the free nucleon-nucleon interactions are first transformed into the effective interactions which are designed to be used in a small model space. Since a major step in these calculations is to derive the reaction matrix which we shall discuss in some detail later, we may generally refer to this type of calculation as the reaction matrix approach for nuclear structure.

In this paper, I shall not attempt to review the various developments which have been carried out in this area in the past few years, since excellent reviews about these developments have already been given by several authors[5-8]. Instead, I shall mainly report on some recent work which is being carried out at Stony Brook. In short, I shall describe how to do nuclear structure calculations using a realistic nucleon-nucleon interaction in the framework of the Green's function method. In Section II, I shall first outline the Green's function formalism. In Section III, I shall discuss how to sum up all the two-body correlations to obtain the reaction matrix. This step is necessitated by the presence of the strong short-ranged repulsions in recent nucleon-nucleon potentials. In the present formalism, the effective nucleon-nucleon interactions in nuclei will be energy dependent, and hence the nuclear energy spectrum should be calculated in a self-consistent way. This will be discussed in Section IV, along with a presentation of some preliminary results of a self-consistent calculation of the energy spectra of O^{18} and F^{18}.

II. THE GREEN'S FUNCTION FORMALISM

A well-known and probably the most powerful method in the theory of many body problems is the Green's function method[9]. As we shall see shortly, a combination of the usual

*Work sponsored by the U.S. Atomic Energy Commission (Contract No. AT (20-1)-4032).

Green's function method and the Brueckner reaction-matrix method appears to provide a very desirable theoretical foundation for doing nuclear structure calculations using a realistic nucleon-nucleon interaction. To illustrate, let us consider the case where our goal is to calculate the energy spectra of the nuclei O^{18} and O^{14}.

We define the two particle Green's function as

$$G(3412,\ T-T') \equiv \langle \Psi_0 | T\{a_4(T)a_3(T)\ a_1^+(T')a_2^+(T')\} | \Psi_0 \rangle \quad (1)$$

where Ψ_0 is the true ground state of O^{16} and let us denote the true ground state energy of O^{16} and $a(T)$ where $N_o=16$, namely the number of nucleons in Ψ_o. $a^+(T)$ and $a(T)$ are the fermion creation and destruction operators for neutrons, and their time dependence is given by

$$a(t) = e^{iHt/\hbar}\ a\ e^{-iHt/\hbar} \quad (2)$$

and similarly for $a^+(T)$. Here

$$H = T + V \quad (3)$$

is the true Hamiltonian. The operator T is the Wick time-ordering operator. We define the energy transform of G(3412, T-T') as

$$G(3412,E) = -\frac{i}{\hbar}\int_{-\infty}^{\infty} G(3412,t)\ e^{iEt/\hbar}\ dt \quad (4)$$

wnere $t = T - T'$.

Let us now insert an unit operator between $a_4(T)a_3(T)$ and $a_1^+(T')a^+_2(T')$ of Eq. (1). We write the unit operator as

$$1 = \sum_n |\Psi_n(N_o+2)\rangle\langle\Psi_n(N_o+2)| + \sum_m |\Psi_m(N_o-2)\rangle\langle\Psi_m(N_o-2)| \quad (5)$$

where Ψ_n and Ψ_m are the true eigenstates of the N_o+2, namely O^{18}, and the N_o-2, namely O^{14}, systems with true energies $E_n(N_o+2)$ and $E_m(N_o-2)$. After inserting such a unit operator in Eq. (1) and performing the energy transform according to Eq. (4), we can write G(3412,E) as

$$G(3412,E) = \sum_n \frac{X_n(34)^+ X_n(12)}{E - [E_n(N_o+2) - E_o(N_o)]} + \sum_m \frac{Y_m(12)^+ Y_m(34)}{E + [E_o(N_o) - E_m(N_o-2)]} \quad (6)$$

where

$$X_n(12) = \langle \Psi_n(N_o+2) | a_1^+ a_2^+ | \Psi_o(N_o) \rangle \quad (7)$$

and

$$Y_m(34) = \langle \Psi_m(N_o-2) | a_4 a_3 | \Psi_o(N_o) \rangle. \quad (8)$$

Hence the poles of the two-particle Green's function G(3412,E) are

$$E_n(N_o+2) - E_o(N_o) \text{ and } E_m(N_o-2) - E_o(N_o)$$

namely the true energies of O^{18} and O^{14} measured from the ground state of O^{16}. Thus in order to calculate the energy spectra of O^{18} and O^{14}, our task is to locate the poles of the two-particle Green's function. We now proceed to locate these poles.

We shall first make a perturbation expansion of the Green's function similar to the propagator expansion in field theory.[10] Recall that $\Psi_o(N_o)$ is the true ground state of O^{16}. Let us choose the unperturbed ground state $|o\rangle$ of O^{16} in the following way. We write the true Hamiltonian as

$$H = T + V = (T + U) + (V - U) \equiv H_o + H_1 \quad (9)$$

where U is a one-body potential to be chosen at our disposal. For example, we may choose it as the one-body harmonic oscillator potential

$$U(r) = \tfrac{1}{2} m\omega^2 r^2. \quad (10)$$

We choose to work in a basis defined by ϕ_i with

$$H_o \phi_i = \varepsilon_i \phi_i. \quad (11)$$

Then with U(r) chosen as shown by Eq. (10), the unperturbed ground state $|o\rangle$ of O^{16} will be a shell model state with the proton and neutron 0s and 0p shells filled. Now if the true ground state of O^{16} is not degenerate (which is generally true), we can show that the Green's function G(3412,T - T') of Eq. (1) can be rewritten as

$$G(3412,T-T') = \lim_{\substack{t\to\infty(1-i\delta) \\ t'\to-\infty(1-i\delta)}} [\langle o | T\{U(t-T) a_4(T) a_3(T) U(T-T') a_1^+(T') \times a_2^+(T') U(T'-t')\} | o\rangle / \langle o | U(t-t') | o\rangle] \quad (12)$$

where the operators in the right hand side of Eq. (12) are now in the interaction representation, namely

$$a_k(t) = e^{iH_o t/\hbar} a_k e^{-iH_o t/\hbar} = a_k e^{-i\varepsilon_k t/\hbar} \quad (13)$$

$$a_k^+ = e^{iH_o t/\hbar} a_k^+ e^{-iH_o/\hbar} = a_k^+ e^{i\varepsilon_k t/\hbar} \tag{14}$$

and U(t-t'), the time development operator, is given by

$$U(t-t') = e^{iH_o t/\hbar} e^{-iH(t-t')/\hbar} e^{-iH_o t'/\hbar}. \tag{15}$$

Here we note that we are working in a basis defined by $H_o\phi_i = \varepsilon_i\phi_i$ with H_o defined by Eq.(9).

Now by expanding U in powers of H_1 (H_1 = V - U as given by Eq.(9)), we can rewrite Eq.(12) as the following:

$$G(3421,T-T') = \sum_{n=o}^{\infty} (-i/\hbar)^n \frac{1}{n!} \int_{-\infty(1-i\delta)}^{\infty(1-i\delta)} dt_1 \cdots \int dt_n \langle o|T[a_4(T)a_3(T)$$

$$\times H_1(t_1)\cdots H_1(t_n)a_1^+(T')a_2^+(T')]|o\rangle_L. \tag{16}$$

This is a very desirable result. We note that the subscript L means that we need to retain the linked diagrams only. To explain this, let us first define the diagrammatic representation of V and U in Figure 1. Then by linked diagrams, we

Fig. 1. Diagrammatic representation of V and U.

mean the diagrams which are at least linked to one external line, namely a line which starts with a label 1 or 2 and terminates with label 3 or 4. For example, the diagrams (1), (2), (3) and (4) of Figure 2 are linked to at least one external line and therefore will be retained. But diagram (5) of Figure 2 is unlinked to any external line, and hence will *not* be included in G(3412,T-T'). Because of the limitation of my time, I shall omit the proof of Eq.(16) in this paper. The proof can be carried out in a similar way as the proof of the Goldstone[11] theorem for nuclear matter. Qualitatively we can understand Eq.(16) by noting that the unlinked diagrams contained in the numerator of Eq.(12) are cancelled by the denominator, namely $\langle o|U(t-t')|o\rangle$.

That we need to include only the linked diagrams in Eq.(16) largely simplifies the calculation of the Green's function and therefore helps us greatly in locating the poles of the Green's function. We shall now factorize Eq.(16) in order to obtain the integral equations for G(3412,T-T') and then locate its poles. The following two schemes of approximations will be adopted:

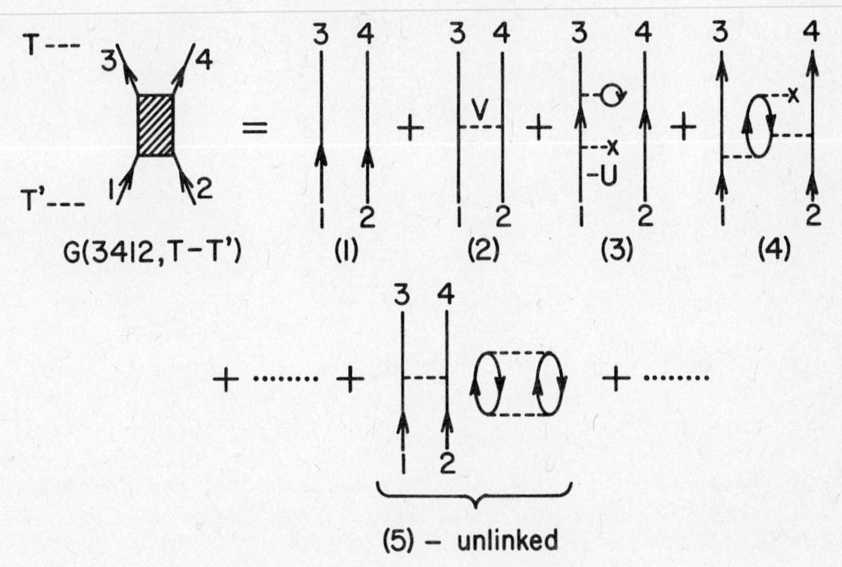

Fig. 2. Diagrammatic expansion of the Green's function G(3412,T-T'). Diagram (5) will *not* be included in the expansion.

APPROXIMATION I

In this approximation, we shall include only the so-called forward going diagrams. The definition of these diagrams will become clear as we proceed a little further. In the simplest case, for example, we shall approximate Eq. (16) by

$$G(3412,T-T') = \delta_{13}\delta_{24}g_1(T-T')g_2(T-T') + \int dx \sum_{56} g_3(T-x)g_4(T-x)(-i/\hbar)\langle 34|V|56\rangle\, G(5612,x-T') \quad (17)$$

where δ_{ij} is the Kronecker delta function, and $g_i(t-t')$ is the free propagator given by

$$\begin{aligned} g_i(t-t') &= e^{i\varepsilon_i(t-t')/\hbar}\theta(t-t') \text{ for } i = \text{particle} \\ &= -e^{i\ {}_i(t'-t)/\hbar}\theta(t'-t) \text{ for } i = \text{hole} \end{aligned} \quad (18)$$

and $\theta(t)$ is the standard step function. $\langle 34|V|56\rangle$ is the bare nucleon-nucleon interaction vertex.

Diagrammatic, Eq. (17) corresponds to the integral equation shown in Figure 3. The energy transform of Eq. (17)

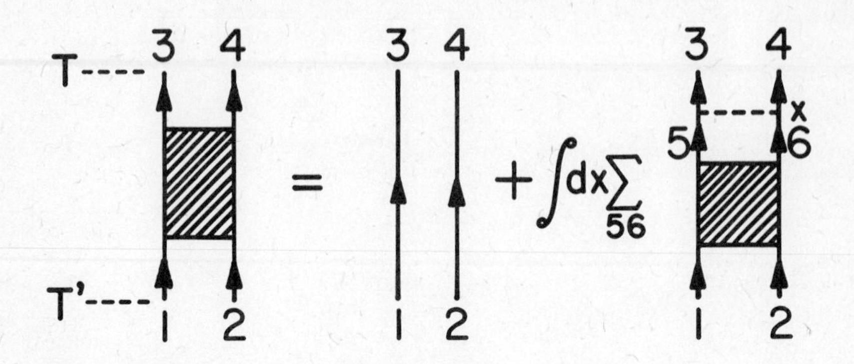

Fig. 3. Diagrammatic representation of Eq. (17). Here we consider 1, 2, 3 and 4 are all particle lines.

is

$$G(3412,E) = \delta_{13}\delta_{24}\frac{1}{E-(\varepsilon_3+\varepsilon_4)} + \sum_{56}\frac{\langle 34|V|56\rangle\, G(5612,E)}{E-(\varepsilon_3+\varepsilon_4)} \tag{18-1}$$

Thus the poles of G(3412),E) are given by the eigenvalues of the secular equation

$$\sum_{56}[(\varepsilon_3 + \varepsilon_4)\, I + \langle 34|V|56\rangle]X_{56} = E\, X_{34} \tag{19}$$

where I is the unit matrix. A more general scheme for approximating Eq. (17) would be the scheme outlined in Figure 4, where Γ_i and M are respectively the irreducible one-body and two-body vertex functions. The diagrams which contribute to Γ_i and M are shown in Figure 5. According to this scheme, the secular equation becomes

$$\sum_{56}[(\varepsilon_3 + \Gamma_3 + \varepsilon_4 + \Gamma_4)\, I + M(3456)]\, X_{56} = EX_{34} \tag{20}$$

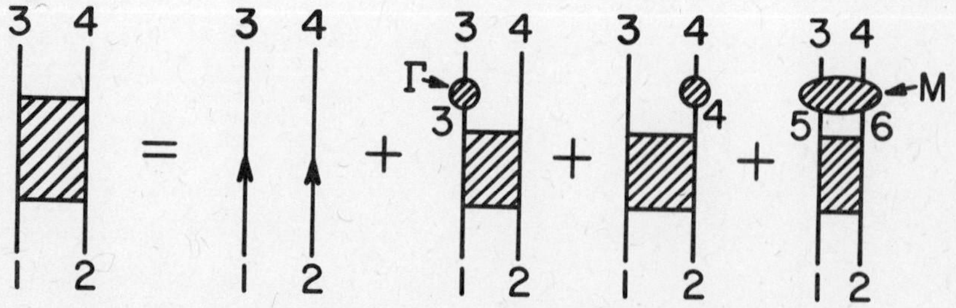

Fig. 4. A more general approximation for Eq. (17).

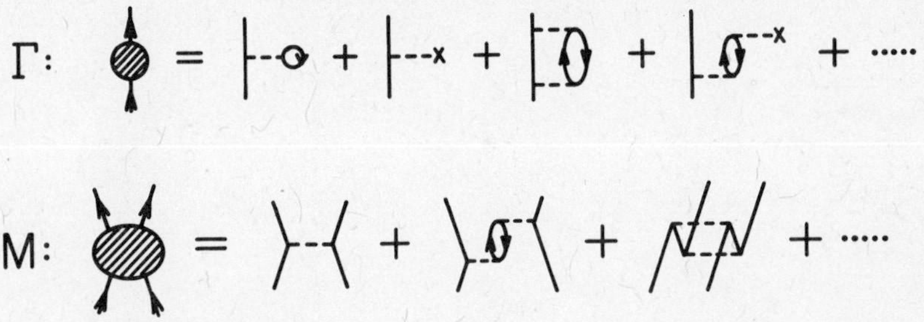

Fig. 5. Irreducible diagrams which contribute to Γ and M.

We see that Eq. (20) differs from Eq. (19) only in replacing ε_i and V of Eq. (19) by $\varepsilon_i+\Gamma_i$ and M. The above equation is now rather similar to the usual shell model secular equation, if we identify $\varepsilon_i+\Gamma_i$ as the experimental single-particle energies and M as the effective interaction. In the present formalism, however, we can systematically calculate Γ_i and M from a given nucleon-nucleon interaction. The various problems involved in the calculation of Γ_i and M will be discussed later in Sections III and IV.

We note that the eigenvalues E of Eq. (20) will correspond to the energies of O^{18} when 1, 2, --- 6 are particle labels, and will correspond to the energies of O^{14} measured from the ground state energy of O^{16} when 1, 2, --- 6 are hole labels. For the latter case, the external lines of Γ_i and M are of course all hole lines.

APPROXIMATION II

In this approximation, we shall include both the forward-going diagrams which are included in Approximation I and the "backward-going" diagrams, and in so doing we shall obtain a coupled equation which will yield the energies of O^{18} and O^{14} at the same time. This approximation can be most conveniently described by writing down the integral equation in a diagrammatic form as shown by Figure 6, where $\overline{G}$(3412, T-T') is just a special case of G(3412, T-T') defined by Eq. (1), namely, we denote G as $\overline{G}$ when 1 and 2 refer to particle labels, and 3 and 4 refer to hole labels, or vice versa. Then similar to what we did for Approximation I, we will obtain the following coupled secular equation:

$$\left\{\begin{aligned} &(\varepsilon_3+\Gamma_3+\varepsilon_4+\Gamma_4)X_{34}+\sum_{56}M(3456)X_{56}+\sum_{\underline{56}}K(34\underline{56})Y_{\underline{56}} = EX_{34} \\ &-\sum_{56}K(\underline{34}56)^*X_{56}-(\varepsilon_{\underline{3}}+\Gamma_{\underline{3}}+\varepsilon_{\underline{4}}+\Gamma_{\underline{4}})Y_{\underline{34}}-\sum M(\underline{3456})^*Y_{\underline{56}} = EY_{\underline{34}} \end{aligned}\right. \tag{21}$$

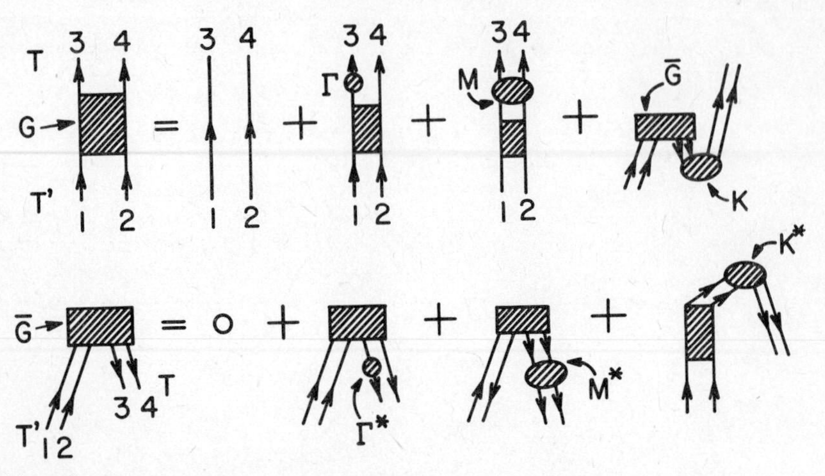

Fig. 6. Diagrammatic representation of Approximation II. For simplicity the diagrams with Γ attached to line 4 are suppressed.

where 1, 2, --- refer to particle labels, and $\underline{1}$, $\underline{2}$ --- refer to hole labels. We may refer to K as the ground state correlation vertex. The irreducible diagrams which contribute to K are shown in Figure 7. The eigenvalues E of Eq. (21) will correspond to the energies of O^{18} and O^{14} measured from the ground state of O^{16}. For the case of O^{18}, the amplitudes X and Y correspond to

$$X_{34} \rightarrow \langle\Psi(O^{18})|a_3{+}a_4{+}|\Psi_o(O^{16})\rangle$$
$$Y_{34} \rightarrow \langle\Psi(O^{18})|a_{\underline{3}}{+}a_{\underline{4}}{+}|\Psi_o(O^{16})\rangle \tag{22}$$

while for O^{14}, they correspond to

$$X_{34} \rightarrow \langle\Psi(O^{14})|a_4 a_3|\Psi_o(O^{16})\rangle$$
$$Y_{34} \rightarrow \langle\Psi(O^{14})|a_{\underline{4}} a_{\underline{3}}|\Psi_o(O^{16})\rangle \tag{23}$$

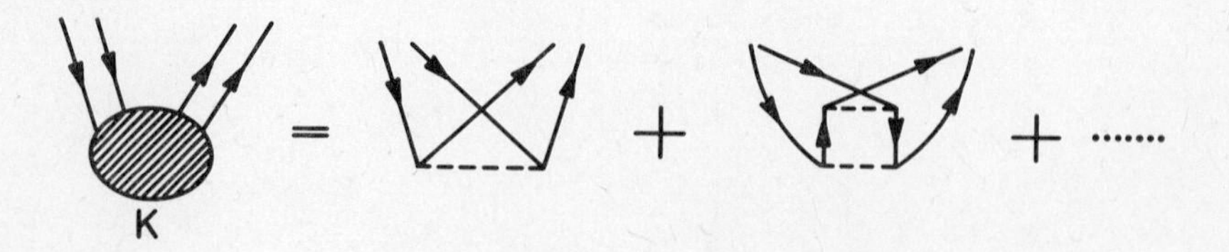

Fig. 7. Irreducible diagrams which contribute to K.

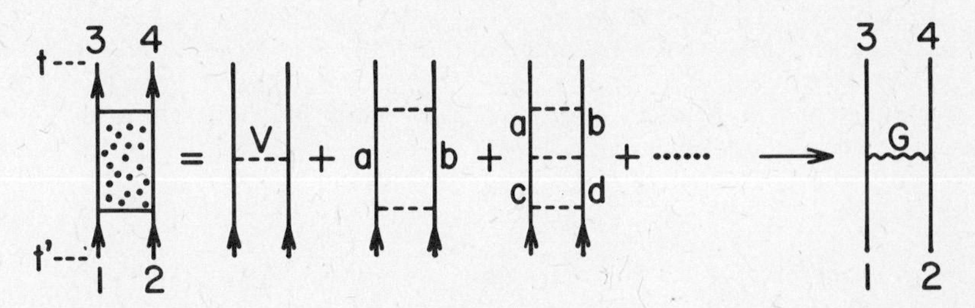

Fig. 8. Diagrammatic representation of $\langle 34|G(t-t')|12\rangle$ or $\langle 34|G(\omega)|12\rangle$.

the physical implications of the above procedure, we should, of course, first define the G matrix. Diagrammatic, it corresponds to the sum of the series of the V interactions shown in Figure 8. As shown, we see that G is a "finite-time" interaction, namely all the V interactions take place within the time interval (t-t'). For the simplicity in drawing diagrams, we shall, however, just draw a wavy line to represent G as shown by the last diagram in Figure 8, although we should remember that this does not mean that G is an instantaneous interaction. The integral equation which defines G is

$$\langle 34|G(E)|12\rangle = \langle 34|V|12\rangle + \sum_{abcd} [1 - C(ab)][1 - C(cd)] \times \langle 34|V|ab\rangle\langle ab|\frac{1}{E-(T+U)}|cd\rangle\langle cd|G(E)|12\rangle \tag{26}$$

where the C's are constants equal to 1 or 0. It is easily seen why we need them in the above equation. For example, if we are solving Eq.(20) within the model space of the 2s-1d shell, then to avoid double counting we must have $C(ab) = 1$ when both a and b belong to the 2s-1d shell. The other purpose of C(ab) is to exclude the filled orbits from the intermediate states of $G(\omega)$. For example, when we solve Eq.(20) for O^{18}, we should set $C(ab) = 1$ if either a or (and) b belongs to the 1s and 1p shells since they are filled by the nucleons of O^{16}*. We shall consider two kinds of choices for C(ab):

(a) The (n_a, n_b, n_c) scheme. This is explained in

*In the theory of many-body problems, we may ignore the exclusion principle if all the direct and exchange diagrams are included. It is, however, not practical to include all the exchange diagrams where an intermediate particle line of $G(\omega)$ exchanges with a core particle line. Hence, we choose to use C(ab) to exclude the filled states from the intermediate states of $G(\omega)$.

Figure 9.

(b) The ρ_{max} scheme, namely the Eden-Emery[13] choice where $C(ab) = 1$ if $(2n_a + \ell_a + 2n_b + \ell_b) \leq \rho_{max}$ (27)

$= 0$ otherwise,

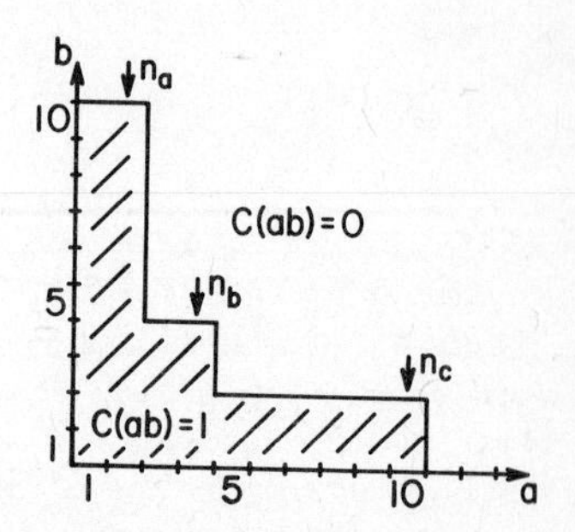

Fig. 9. A choice of C(ab). Here C(ab) equals 1 inside the shaded area, and 0 everywhere else. We shall denote this choice by the values of n_a, n_b, n_c. For example, the present choice corresponds to $(n_a, n_b, n_c) = (2,4,10)$. The numerals 1,2 represent the oscillator orbits, as shown by Table I. The C's are symmetric, namely $C(ab) = C(ba)$.

where ρ_{max} is an integer. It is clearly seen that the choice of C(ab) can be diagram-dependent. Namely for some irreducible diagrams in the vertex functions we need to exclude both the intermediate states which will cause double-counting and those which correspond to the filled states. But for some other diagrams, we don't need to worry about the problem of double-counting.

TABLE I

Correspondence between the orbit numbers shown in the left column and the n, ℓ quantum numbers.

	nℓ		nℓ		nℓ		nℓ		nℓ
1	os	6	1p	11	1f	16	3s	21	ok
2	0p	7	0g	12	2p	17	oj	22	1i
3	0d	8	1d	13	0i	18	1h	23	2g
4	1s	9	2s	14	1g	19	2f	24	3d
5	0f	10	0h	15	2d	20	3p	25	4s

We shall use the free-particle spectrum for the intermediate states of Eq.(26). Namely, for the intermediate states we shall include the *diagonal* U insertions to all orders such as diagram (a) of Figure 10. But we shall *not* include any self-energy V insertions for the intermediate states, such as diagram (b) of Figure 10. There has been

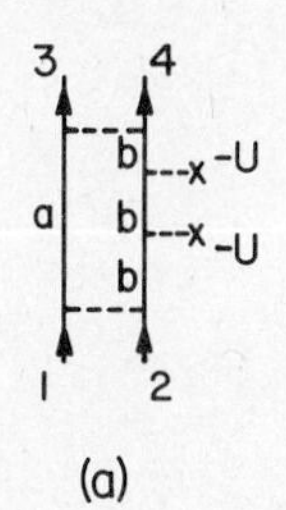

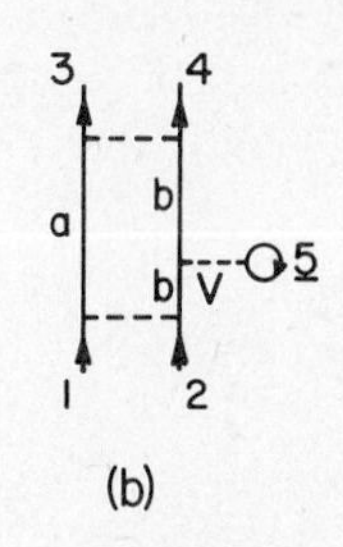

Fig. 10. Diagrams corresponding to the U and V insertions to the intermediate states of $G(\omega)$.

much discussion with regard to the use of the free-particle spectrum for the intermediate states[5,6,7]. Generally speaking, the idea for doing so is the following. The nucleus is a low density system, thus when a pair of nucleons interact with each other, the rest of the nucleons in the nucleus seldom participate. Hence, the pair of nucleons interact with each other as if they were free particles, except for the blocking of certain states and the off-energy shell effects caused by the presence of other nucleons. We should put back the U insertions, since originally there is no U in the Hamiltonian. Recall that in Eq. (9) we add and then subtract U. As for the V insertions, Bethe[14] has shown for nuclear matter that one should sum the three-body cluster diagrams to all order, and then the net contribution from the three-body clusters is nearly zero. No quantitative study for the three-body clusters in nuclei has been done, and before such a study is carried out, it is probably best to just leave out the V insertions for the intermediate states of $G(\omega)$. All the *diagonal* U insertions can be summed up to cancel the U in the denominator of Eq. (26), namely Eq. (26) now becomes

$$\langle 34|G(E)|12\rangle = \langle 34|V|12\rangle + \sum_{abcd} [1-C(ab)][1-C(cd)] \times \langle 34|V|ab\rangle\langle ab|\frac{1}{E-T}|cd\rangle\langle cd|G(E)|12\rangle \; . \tag{28}$$

We would like to solve the G-matrix equation in the momentum representation. The general idea is that we shall write the G-matrix equation in a form like

$$\int_0^\infty q^2 dq K(pq) \; \langle q|G(E)|r\rangle = \langle p|V|r\rangle \tag{29}$$

where K and V are known and p, q, and r refer to the relative momentum of the pair of interacting nucleons. Then since K(pq) is generally a smooth function in the momentum variables and it goes to zero sufficiently fast as q and r approach infinity, we can use the Gauss quadratures[15] to approximate Eq.(29) by a finite matrix equation, namely

$$\sum_{i=1}^{N} W(p_i)p_i^2 K(p_m\ p_i)\ \langle p_i|G(E)|p_n\rangle = \langle p_m|V|p_n\rangle \tag{30}$$

where N is the number of Gauss points and the W's are the Gauss weights. Usually $N \approx 20$ is quite adequate. Then $\langle P_m|G(E)|P_n\rangle$ can be obtained easily by solving Eq. (30). The above method has been used by Brown, Jackson and Kuo[16] in the calculations of phase shifts and the nuclear matter reaction matrix elements, and was found to be very satisfactory. Thus we feel confident in using this method to solve the shell model reaction matrix elements.

To cast Eq. (28) into the form of Eq. (29), we choose to write Eq. (28) in the basis of

$$|NL, k\ell sj, J\rangle \tag{31}$$

where NL represent the oscillator center-of-mass function $\phi_{NL}(R)$ for the pair of interacting nucleons. K refers to the relative momentum. ℓ, s and J denote the relative angular momentum, total spin and the total angular momentum. $\vec{j} = \vec{\ell}+\vec{s}$. Let us write a projection operator.

$$P = \sum_{ab} C(ab)\ |ab\rangle\ \langle ab| \tag{32}$$

where C equals to 0 or 1 according to Figure 9 or Eq. (27). Then, symbolically we can write Eq. (28) as

$$G(E) = V + V\ (1-P)\ \frac{1}{E-T}\ (1-P)\ G(E) \tag{33}$$

In j-j coupling, P should be written as

$$P = \sum_{56} C(n_5\ell_5 j_5,\ n_6\ell_6 j_6)\ \ |(n\ell j)_5 (n\ell j)_6\rangle\langle (n\ell j)_5 (n\ell j)_6| \tag{34}$$

where C(---) equals to 1 or 0 according to which states we want to exclude from G(E). If the $j = \ell \pm \frac{1}{2}$ levels are both excluded or included, then we have

$$C(n_5\ell_5 j_5,\ n_6\ell_6 j_6) = C(n_5\ell_5,\ n_6\ell_6) = 0 \text{ or } 1 \tag{35}$$

and can write P in the λ-s coupling scheme. After performing the Moshinsky transformation, P becomes

$$P = \sum_{\substack{n_5 \ell_5 n_6 \ell_6 \lambda S \\ NLn\ell n'\ell' jj' \\ N'L'}} C(n_5\ell_5, n_6\ell_6)\, M_\lambda(n_5\ell_5 n_6\ell_6 NLn\ell)\, M_\lambda(n_5\ell_5 n_6\ell_6 N'L'n'\ell')$$
$$\times <(L\ell)\lambda SJ|L(\ell S)jJ><(L'\ell')\lambda SJ|L'(\ell S')j'J>$$
$$\times |NL, n\ell Sj, J><N'L', n'\ell' Sj', J| \quad (36)$$

where M_λ(----) is the Moshinsky transformation brackets and <--|--> is the coefficient corresponding to the angular momentum recoupling from

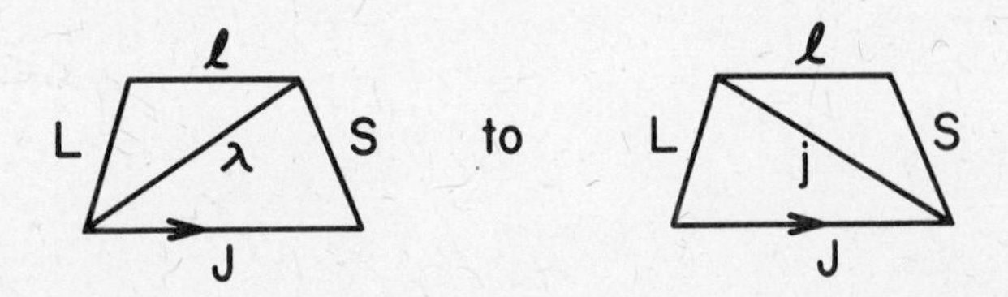

Our purpose has been to write Eq. (33) in the basis of $|NL, k\ell sj, J>$. This can now be accomplished if we make a Fourier transformation of Eq. (26), namely replace n and n' by k and k'. But this will lead to a very complicated and coupled equation, mainly because P is not diagonal in the $|NL, k\ell sj, J>$ representation although it is diagonal in the j-j representation of Eq. (34).

We shall discuss under what circumstances the form of P can be simplified. If we use the Eden-Emery choice of C(ab) as shown by Eq. (27), then we can make use of the closure property of the Moshinsky transformation brackets and obtain a very desirable form of P, namely

$$P_E = \sum_{NLn} C(2N+L+2n+\ell)\ |NL, n\ell Sj, J> <NL, n\ell Sj, J| \quad (37)$$

where

$$C(2N+L+2n+\ell) = 1, \text{ if } (2N+L+2n+\ell) \leq \rho_{max} \quad (38)$$
$$= 0, \text{ otherwise.}$$

For a more general choice of C(ab) as shown by Figure 9, it will be very difficult to solve Eq. (33) if we do not make some approximation of P. We shall adopt the following approximation which is similar to the "angle-averaged" approximation of Wong[7]. Namely, we take P to be diagonal in ℓ and L, and average the Moshinsky brackets over λ. Then we can write P of Eq. (36) as

$$P = \sum_{\substack{n_5\ell_5n_6\ell_6 \\ NnN'n'L\ell}} C(n_5\ell_5n_6\ell_6) \left\{ \sum_\lambda \frac{2\lambda+1}{(2\ell+1)(2L+1)} M_\lambda(n_5\ell_5n_6\ell_6NLn\ell) \times M_\lambda(n_5\ell_5n_6\ell_6N'Ln'\ell) \right\} |NL,n\ell Sj,J\rangle\langle N'L,n\ell' Sj,J|. \tag{39}$$

We further take P to be diagonal in N (then it must be also diagonal in n) and denote it as P_D. Then P_D is simply

$$P_D = \sum_{\substack{n_5\ell_5n_6\ell_6 \\ NLn\ell}} A(n_5\ell_5n_6\ell_6, NL\,n\ell)|NL,n\ell Sj,J\rangle\langle NL,n\ell Sj,J| \tag{40}$$

with

$$A(n_5\ell_5n_6\ell_6, NLn\ell) = C(n_5\ell_5n_6\ell_6) \sum_\lambda \frac{2\lambda+1}{(2\ell+1)(2L+1)} \times M_\lambda(n_5\ell_5n_6\ell_6NLn\ell)^2. \tag{41}$$

Here we see that A(---) corresponds to the probability of excluding $|NL, n\ell sj,J\rangle$ from the intermediate states of $G(\omega)$ when the state $|n_5\ell_5n_6\ell_6\rangle$ is definitely excluded in the $n_a\ell_a n_b\ell_b\rangle$ representation. The matrix element of P_D in the $|NL,k\ell sj,J\rangle$ basis is

$$\langle NL,\sqrt{\tfrac{2}{\pi}}\, j_\ell(kr), \ell Sj,J|\, P_D\, |NL,\sqrt{\tfrac{2}{\pi}}\, j_\ell(k'r),\ell Sj,J\rangle$$

$$= \sum_{\substack{n_5\ell_5n_6\ell_6 \\ n}} A(n_5\ell_5n_6\ell_6, NLn\ell) \frac{P_{n\ell}(k)\, P_{n\ell}(k')}{kk'} \tag{42}$$

$$\equiv \langle k,NL\ell|P_D|k',NL\ell\rangle$$

where j_ℓ is the spherical Bessel function, and $P_{n\ell}(k)$ is the k-space harmonic oscillator wave function, namely

$$\langle \vec{k}_\delta|\, \frac{R_{n\ell}(r)}{r}\, Y_{\ell m}(\hat{r})\rangle \equiv \frac{P_{n\ell}(k)}{k}\, Y_{\ell m}(\hat{k}) \tag{44}$$

where the oscillator constant ν_k for $R_{n\ell}(k)$ is related to ν_r for $R_{n\ell}(r)$ by $\nu_k = (\nu_r)^{-1}$. $\vec{k}_\delta$ represents the δ-function normalized plane wave.

Knowing now the matrix elements of P_D in k-space, we are now ready to write Eq. (33) into the form of Eq. (29). Before doing so, we should first write (E-T) in Eq. (33) in a more general form. Eq. (33) is for the G-matrix equation where there are only two particle lines. When there are other lines present, the energy denominator in Eq. (33) will take a different form. Consider the example shown in Figure 11. A term which is contained in the upper G-matrix is shown in the upper right corner of the figure. As shown, the energy denominator for $\langle h4|G(E)|p2\rangle$ is clearly

$$E - (T_a+T_b+h_5+h_6+h_3-h_h)$$

where a and b are the intermediate states in the G-matrix, and T_a and T_b are their kinetic energies. That we have T_a and T_b is because we choose to use the free-particle spectrum. The h's are the single-particle energies of the other particle and hole lines. They would be just T_5, T_6---, if we should choose to use the free particle spectrum for them also. We write them as h's, because we shall use a G-matrix Hartree-Fock spectrum for them. Thus, for a general case, (E-T) of Eq. (33) should be replaced by

$$E - (T_a+T_b + \sum_p h_p - \sum_h h_h)$$

$$= E - (k^2+\frac{K^2}{4} + \sum_p h_p - \sum_h h_h) = -k^2 - \frac{K^2}{4} + \omega \qquad (45)$$

where k and K are respectively the relative and center-of-mass momentum of particles a and b, and

$$\omega = (E+\sum_h h_h - \sum_p h_p) \qquad (46)$$

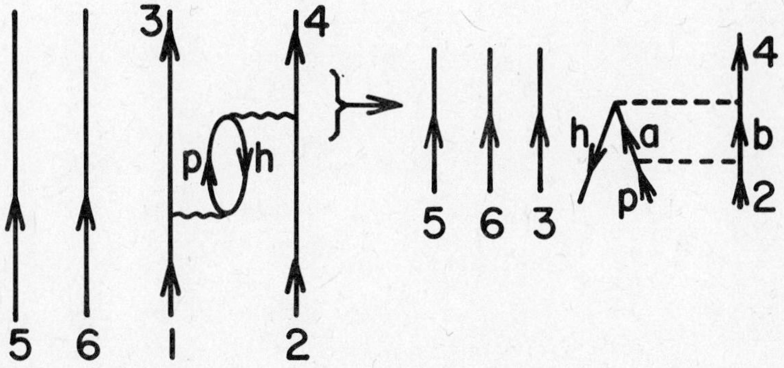

Fig. 11. A typical diagram which belongs to the vertex function of a four-particle Green's function. Each wavy line represents a G-matrix.

Then Eq. (33) can be rewritten as

$$G(\omega) = V + V\,(1-P)\,\frac{1}{\omega-\frac{K^2}{4}-k^2}\,(1-P)\,G(\omega) \tag{47}$$

If in addition to taking P to be diagonal, namely replacing P by P_D, we also consider K^2 to be diagonal in N, we can write Eq. (47) in the following form:

$$\begin{aligned}
\langle k'|G_{\ell\ell}(\omega,NL)|k''\rangle &= \langle k'|V_{\ell\ell}|k''\rangle \\
&+\{\tfrac{2}{\pi}\int k^2dk\langle k'|V_{\ell\ell}|k\rangle E(k,NL\omega) \\
&+\tfrac{2}{\pi}\int k^2dk\int q^2dq\langle k'|V_{\ell\ell}|q\rangle \quad \langle q,NL\ell|P_D|k,NL\ell\rangle[E(k,NL\omega)+E(q,NL\omega)] \\
&-\tfrac{2}{\pi}\int k^2dk\int q^2dq\int p^2dp\ \langle k'|V_{\ell\ell}|q\rangle \quad \langle q,NL\ell|P_D|p,NL\ell\rangle \\
&\times\langle p,NL\ell|P_D|k,NL\ell\rangle E(p,NL\omega)\}\ \langle k|G_{\ell\ell}(\omega,NL)|k''\rangle
\end{aligned} \tag{48}$$

$$\text{where } E(k,NL\ ) = \int dk\,\frac{P_{NL}(K)^2}{\omega-k^2-\frac{K^2}{4}} \tag{49}$$

Here $P_{NL}(K)$ is the harmonic oscillator wave function for the center-of-mass momentum, defined similarly as $P_{n\ell}(k)$ of Eq. (43). Eq. (48) is now in the form of Eq. (29), and can be solved straightforwardly on a computer. After Eq. (48) is solved, the G-matrix elements for the harmonic oscillator wave functions are obtained simply by carrying out the following integral:

$$\langle n\ell|G_{\ell\ell}(\omega,NL)|n\ell\rangle = \frac{2}{\pi}\int kdkk'dk'P_{n\ell}(k')\ \times\langle k|G_{\ell\ell}(\omega,NL)|k'\rangle \tag{50}$$

The above is for the case of a single ℓ channel. For the coupled-channel case where ℓ and ℓ' are coupled such as the $^3S_1$ - $^3D_1$ channel, we can easily generalize Eq. (48) to obtain an equation for

$$\langle k'|G_{\ell\ell'}(\omega,NL)|k''\rangle$$

In obtaining Eq. (48), we have used the diagonal approximation for P and $(+\omega-k^2-\frac{K^2}{4})^{-1}$. It is not clear how good this approximation is. Although we have looked at some off-diagonal matrix elements of P and found them to be generally small compared with the diagonal matrix elements of P, this is, however, not adequate in insuring that the diagonal approximation is a good approximation. An investigation of this problem is being carried out by K. Ratcliff, S.F. Tsai and myself.

We now consider another approximation for Eq. (47), namely in addition to replace P by P_D, we replace $K^2/_4$ by a constant which equals to its average value

$$\langle\frac{K^2}{4}\rangle_{NL} = 1/2\ (2N+L+\frac{3}{2})\hbar\omega \tag{51}$$

Then Eq. (47) becomes

$$G(\gamma^2) = V+V(1-P_D)\ \frac{1}{-k^2-\gamma^2}\ (1-P_D)\ G(\gamma^2) \tag{52}$$

where

$$\gamma^2 = -\ \omega + \langle\frac{k^2}{4}\rangle_{NL} \tag{53}$$

Numerically, we have found that the matrix elements given by Eq. (52) are very similar to those given by

$$G(\gamma^2) = V+V(1-P_D)\ \frac{1}{-k^2-\gamma^2}\ G(\gamma^2) \tag{54}$$

Similar to what we did for Eq. (47), we can easily write Eq. (52) or (54) in a form like Eq. (48), and then solve them according to the method outlined by Eqs. (29) and (30).

IV. SELF-CONSISTENT CALCULATIONS

Using the reaction matrix approach outlined in the previous section, the vertex functions Γ, M and K of Eq. (20) and (21) will be a function of E. For example, the contribution from the irreducible diagram shown in Figure 12 is, aside from the angular momentum recoupling coefficients,

$$\frac{(-1)^2\langle h4|G(\omega_1)|p^2\rangle\ \langle 3p|G(\omega_2)|1h}{E-(h_p+h_3+h_2-h_h)}$$

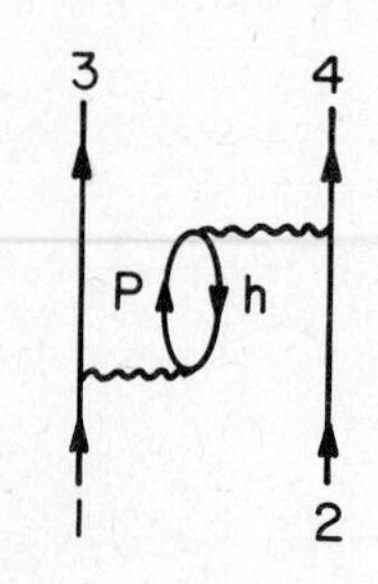

Fig. 12. A diagram which contributes to the vertex function of M of Eq.(20) and Eq.(21). Each wavy line represents a G-matrix.

with

$$\omega_1 = E-(h_3-h_h)$$

$$\omega_2 = E-(h_2-h_h)$$

Thus the secular equations Eq. (20) and Eq. (21) are in fact of the general form

$$H(E)\Psi = E\Psi \tag{55}$$

and will be solved self-consistently. Namely, we may first assume an E' and use it to calculate H(E') and E. Then repeat the process until E' equals to E. If the dependence of H on E is fairly weak, then the self-consistence procedure can be carried out easily using a graphical method which we shall discuss later.

We shall now report some preliminary results we have made for the nuclei O^{18} and F^{18}. We shall use the Reid[12] soft core nucleon-nucleon potential. As we saw in Section II, the G-matrix equation can be solved rather easily in the momentum space representation if we make the diagonal approximations. But, conceptually, it is not very clear which of the three equations, namely Eq. (48), Eq. (52), and Eq. (54), provides the best approximation for the original G-matrix equation. So far, our numerical calculations have been done mainly with Eq. (54). This is just for simplicity in numerical calculations. We have found numerically that the matrix elements given by Eq. (54) are only slightly non-Hermitian. We shall now discuss our results in the following three categories:

IV.1 Dependence of $G(\gamma^2)$ on the choice of P.

Recall that the projection operator P is defined by Eq. (32) and we have used two schemes to define the coefficients C(ab), namely the (n_a, n_b, n_c) scheme of Figure 9 and the Eden-Emery scheme of Eq. (27). For the calculation of O^{18} and F^{18}, we should use $n_a=2$ and $n_c=\infty$ in the (n_a, n_b, n_c) scheme. Clearly, it is not practical to use $n_c=\infty$ in calculation. As shown by Table II, the convergence with respect to n_c is very desirable for the ^1S_o matrix elements. Only the $^3S_1$ - $^3S_1$ matrix elements do not converge desirably as n_c increases. The ρ_{max} matrix elements shown in the table were calculated with P set to zero in the G-matrix equation.

Generally speaking, the effect of P is important only for the ^3S_o and $^3S_1$ - $^3D_1$ matrix elements, especially for the latter. As shown by Table II, the (n_a, n_b, n_c) matrix elements differ from the P=o matrix elements by about 2 MeV for the $^3S_1$ - $^3S_1$ case. As for the variation of the matrix elements shown in Table II with respect to n_b, we find that only the $^3S_1$ - $^3S_1$ matrix elements are rather sensitive to n_b. For (n_a, n_b, n_c) = (2,4,20), (2,9,20) and (2,12,20) the values for the $^3S_1$ - $^3S_1$ matrix elements are -6.810, -6.270 and -5.895 MeV. Note that the corresponding values for (n_a, n_b, n_c) = (2,6,20) are listed in Table II.

We have found that for the $^1P_1$, ^3P_o, $^3P_1$, $^3P_2$-$^3F_2$, $^1D_2$ and $^3D_2$ channels of the Reid potential, the effect of P is essentially negligible.

IV.2 Dependence of the G-matrix on γ^2,ω and NL.

For the radial matrix elements $\langle n\ell|G(\gamma^2,NL)|n'\ell'\rangle$, we have found that their dependence on γ^2 is very smooth, and is almost linear. The variation with γ^2 is fairly strong only for the $^1S_1$ - $^3S_1$ matrix elements, as shown by Figure 13. The variation of the $^1S_1$ - $^3S_1$ matrix elements with respect to γ^2 is much less, as shown by Figure 14. For all the other partial waves of the Reid potential, the variation with γ^2 is even less. As for the variation with NL, we have found that the matrix elements are very nearly equal to each other if they have the same (2N+L) values. This is of course so, if we use the Eden-Emery choice of P, but it is not clear why the matrix elements of same (2N+L) values are nearly equal to each other when we use the (n_a, n_b, n_c) scheme for P, such

TABLE II

Dependence of the matrix elements $\langle n\ell|G(\gamma^2,NL)|n'\ell'\rangle$ on the choice of P. The matrix elements were calculated with Eq. (54), with $\gamma^2 = 2\hbar^2/M_N$ where M_N is the nucleon mass and $\hbar\omega$ = 14 MeV. All matrix elements are in units of MeV. The Reid potential is used.

	$\ell\ell'$ n n'N L	(n_a, n_b, n_c) (2,6,6)	(2,6,9)	(2,6,12)	(2,6,16)	(2,6,20)	$\rho_{max}=6$	P = 0
$^1S_0$	0 0 0 0 0 0	-6.186	-6.178	-6.175	-6.175	-6.175	-6.169	-6.890
$^3S_1 - {}^3D_1$	0 0 0 0 0 0	-6.860	-6.795	-6.700	-6.684	-6.652	-5.499	-8.786
	0 2 0 0	-5.839	-5.842	-5.849	-5.851	-5.852	-6.005	-6.410
	2 0 0 0	-5.836	-5.839	-5.846	-5.847	-5.849	-5.999	-6.410
	2 2 0 0	1.350	1.353	1.355	1.355	1.355	1.381	.860

as those of Table II. As shown by Figures 13 and 14, the dependence on (N,L) is strong when γ^2 is small, and becomes rather weak when γ^2 is large. When γ^2 is very small, say in the vicinity of 0.1, the variations of the radial matrix elements with respect to γ^2 are not as smooth and regular as those shown by Figures 13 and 14.

The particle-particle G-matrix elements $\langle 12|G(\omega)|34\rangle$ are calculated by taking appropriate summation[1] of the above radial matrix elements, with the value of γ^2 related

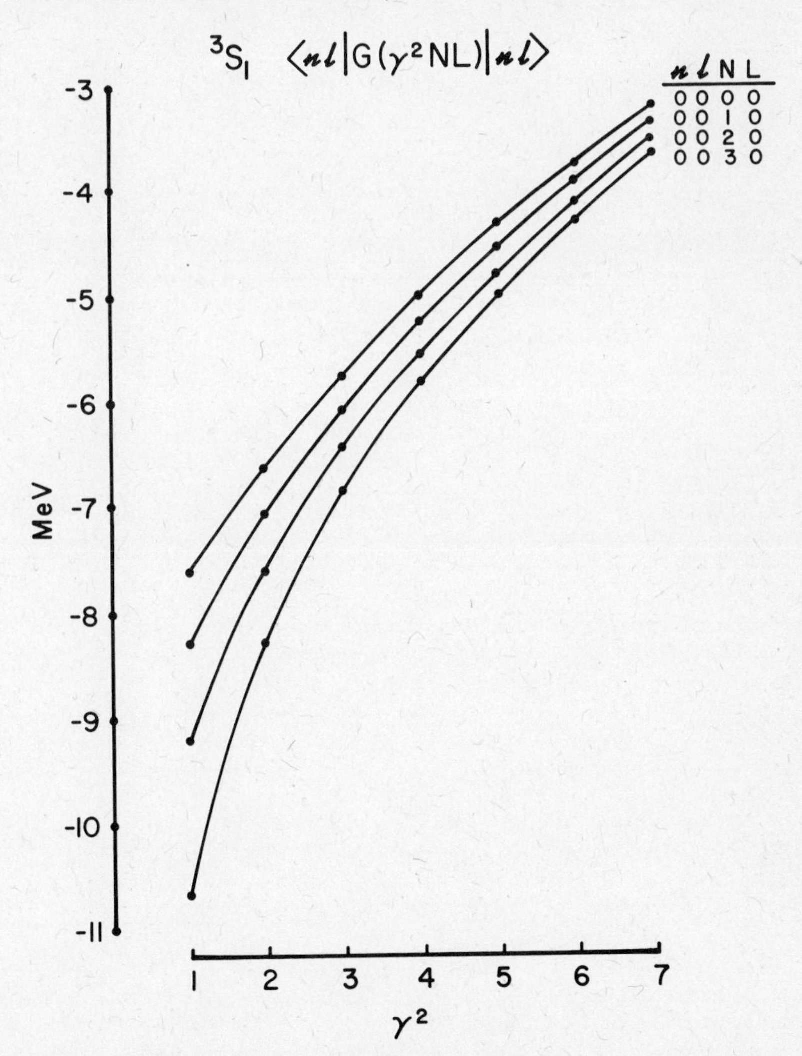

Fig. 13. Dependence of the $^3S_1-{}^3S_1$ matrix elements on γ^2. The matrix elements were calculated with $\hbar\omega \equiv 14$ and with $(n_a,n_b,n_c) = (2,6,20)$.

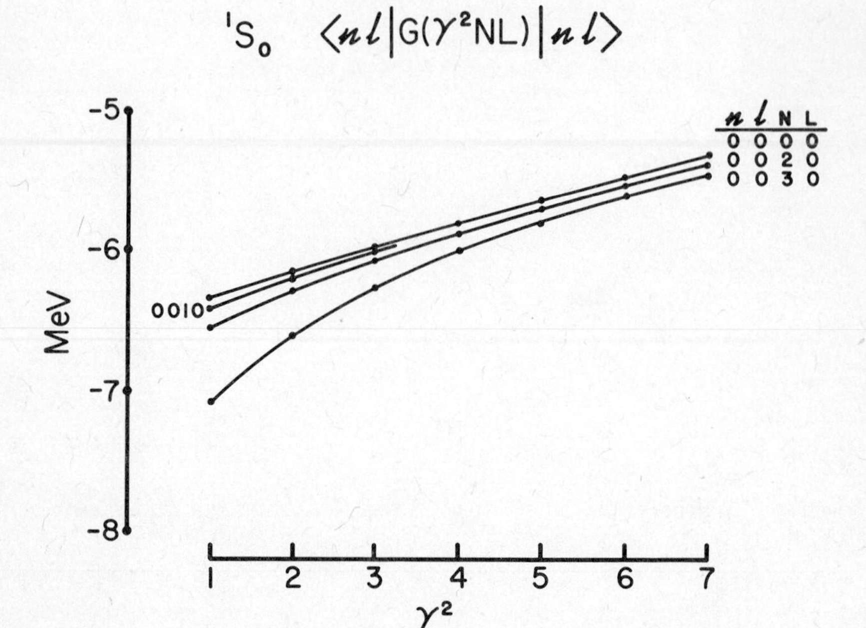

Fig. 14. Dependence of the $^1S_0$ matrix elements on γ^2. The matrix elements were calculated with $\hbar\omega = 14$ and (n_a, n_b, n_c) equal to (2,6,20).

to ω according to Eq. (54). Because of the smooth variation of the radial matrix elements with γ^2, we only calculate them for several values of γ^2 and use an interpolation formula to obtain the radial matrix elements for any values of γ^2 which we need in the calculation of $\langle 12|G(\omega)|34\rangle$. $\langle 12|G(\omega)|34\rangle$ is found to vary very smoothly with ω. This is a very desirable result, because it permits us to calculate the G(ω) matrix elements for just a few values of ω and obtain the matrix elements of other values by interpolation.

When we use the (n_a, n_b, n_c) scheme for the P operator, there are several uncertainties in the treatment of P as we discussed in Section II. Thus we feel that it is probably better to just use the Eden-Emery choice where P is diagonal in the $|NL, n\ell Sj, J\rangle$ representation. We have chosen a model space defined by $\rho_{max} = 6$ for the calculation of the nuclei O^{18} and F^{18}. In Table III we tabulate all the 2s-1d shell matrix elements for five values of ω. We see that the variation with ω is indeed smooth and small in general. The variation with ω is rather large only for a few T = 0 matrix elements where the contributions from the $^3S_1$-$^3S_1$ radial matrix elements are large.

TABLE III

Matrix elements <12JT|G(ω)|34JT> for the 2s-1d shell. They were calculated with $\hbar\omega = 14$, the Reid potential and a projection operator defined by $\rho_{max} = 6$. Both ω and the matrix elements are in units of MeV. The labelling of the orbitals are

$$4 \rightarrow 1d_{5/2},\ 5 \rightarrow 2s_{1/2},\ 6 \rightarrow 1d_{3/2}$$

The matrix elements in the last column are taken from Ref. 2.

	ω					
T a b c d J	-4.	-10.	-16.	-40.	-70.	G(HJ)
1 4 4 4 4 0	-1.3629	-1.3368	-1.3109	-1.2129	-1.0989	-1.236
1 4 4 4 4 2	-1.2494	-1.2402	-1.2312	-1.1985	-1.1627	-1.012
1 4 4 4 4 4	-0.5817	-0.5766	-0.5716	-0.5538	-0.5351	- .434
1 4 4 4 5 2	-0.5874	-0.5832	-0.5792	-0.5640	-0.5465	- .562
1 4 4 4 6 2	-0.2338	-0.2329	-0.2319	-0.2274	-0.2201	- .407
1 4 4 4 6 4	-0.9879	-0.9834	-0.9792	-0.9630	-0.9441	-1.047
1 4 4 5 5 0	-0.7030	-0.6946	-0.6866	-0.6558	-0.6201	- .626
1 4 4 5 6 2	-0.5733	-0.5696	-0.5660	-0.5528	-0.5377	- .546
1 4 4 6 6 0	-3.1835	-3.1691	-3.1547	-3.0991	-3.0311	-3.025
1 4 4 6 6 2	-0.5112	-0.5089	-0.5065	-0.4969	-0.4843	- .598
1 4 5 4 5 2	-1.3008	-1.2920	-1.2834	-1.2522	-1.2178	-1.169
1 4 5 4 5 3	-0.5445	-0.5404	-0.5363	-0.5223	-0.5085	- .288
1 4 5 4 6 2	-0.1590	-0.1558	-0.1526	-0.1408	-0.1273	- .183
1 4 5 4 6 3	-0.1867	-0.1866	-0.1866	-0.1870	-0.1882	- .063
1 4 5 5 6 2	-1.3388	-1.3352	-1.3317	-1.3185	-1.3026	-1.449

TABLE III Con'td.

T a b c d J	ω -4.	-10.	-16.	-40.	-70.	G(HJ)
1 4 5 6 6 2	-0.8111	-0.8079	-0.8047	-0.7929	-0.7796	- .750
1 4 6 4 6 1	-0.6703	-0.6622	-0.6542	-0.6254	-0.5949	- .332
1 4 6 4 6 2	-0.6207	-0.6136	-0.6066	-0.5817	-0.5554	- .356
1 4 6 4 6 3	-0.7594	-0.7565	-0.7537	-0.7439	-0.7340	- .403
1 4 6 4 6 4	-2.0636	-2.0518	-2.0403	-1.9983	-1.9513	-2.023
1 4 6 5 6 1	-0.1292	-0.1294	-0.1296	-0.1307	-0.1323	- .168
1 4 6 5 6 2	-0.7193	-0.7171	-0.7149	-0.7068	-0.6976	- .657
1 4 6 6 6 2	-0.7905	-0.7875	-0.7846	-0.7735	-0.7606	- .777
1 5 5 5 5 0	-2.1890	-2.1697	-2.1508	-2.0787	-1.9943	-2.049
1 5 6 6 6 0	-0.5740	-0.5672	-0.5606	-0.5355	-0.5063	- .526
1 5 5 5 6 1	-0.6151	-0.6104	-0.6057	-0.5890	-0.5719	- .328
1 5 5 5 6 2	-0.7543	-0.7469	-0.7397	-0.7139	-0.6860	- .592
1 5 6 6 6 2	0.0061	0.0088	0.0115	0.0214	0.0329	- .041
1 6 6 6 6 0	-0.0633	-0.0431	-0.0230	0.0523	0.1386	- .087
1 6 6 6 6 2	-0.4039	-0.3982	-0.3926	-0.3721	-0.3499	- .281
0 4 4 4 4 1	0.0535	0.1230	0.1909	0.4423	0.7201	-0.296
0 4 4 4 4 3	-0.7906	-0.7514	-0.7133	-0.5736	-0.4208	-0.791
0 4 4 4 4 5	-3.2028	-3.1211	-3.0431	-2.7633	-2.4700	-3.422
0 4 4 4 5 3	-1.1651	-1.1305	-1.0973	-0.9774	-0.8500	-1.238
0 4 4 4 6 1	2.5392	2.4814	2.4250	2.2162	1.9849	2.595
0 4 4 4 6 3	1.1973	1.1780	1.1592	1.0898	1.0133	1.467
0 4 4 5 5 1	-0.0846	-0.0595	-0.0353	0.0550	0.1541	-0.268
0 4 4 5 6 1	0.1688	0.1858	0.2023	0.2627	0.3281	-0.106
0 4 4 6 6 1	1.6975	1.6644	1.6318	1.5103	1.3736	2.097
0 4 4 6 6 3	0.1592	0.1445	0.1303	0.0782	0.0215	0.389

TABLE III Con'td.

T a b c d J	ω -4.	-10.	-16.	-40.	-70.	G(HJ)
0 4 5 4 5 2	-0.6241	-0.5903	-0.5579	-0.4409	-0.3173	-0.533
0 4 5 4 5 3	-2.9153	-2.8486	-2.7849	-2.5563	-2.3163	-3.121
0 4 5 4 6 2	-1.1889	-1.1605	-1.1334	-1.0348	-0.9298	-1.296
0 4 5 4 6 3	0.8436	0.8171	0.7916	0.6997	0.6021	1.014
0 4 5 5 6 2	-2.2351	-2.2076	-2.1813	-2.0870	-1.9881	-2.514
0 4 5 6 6 3	0.2865	0.2800	0.2749	0.2542	0.2321	0.115
0 4 6 4 6 1	-4.2107	-4.0811	-3.9547	-3.4900	-2.9779	-4.331
0 4 6 4 6 2	-3.7622	-3.7039	-3.6473	-3.4420	-3.2205	-3.592
0 4 6 4 6 3	-1.2406	-1.2035	-1.1677	-1.0381	-0.8993	-1.113
0 4 6 4 6 4	-3.7295	-3.6472	-3.5686	-3.2868	-2.9918	-4.158
0 4 6 5 5 1	1.7245	1.6850	1.6473	1.5087	1.3589	1.609
0 4 6 5 6 1	-1.6909	-1.6670	-1.6444	-1.5648	-1.4820	-1.423
0 4 6 5 6 2	-1.4561	-1.4214	-1.3881	-1.2673	-1.1388	-1.594
0 4 6 6 6 1	-0.0628	-0.0418	-0.0214	0.0553	0.1405	-0.110
0 4 6 6 6 3	1.5619	1.5389	1.5171	1.4398	1.3602	1.725
0 5 5 5 5 1	-2.8561	-2.7685	-2.6837	-2.3705	-2.0268	-3.008
0 5 5 5 6 1	0.2194	0.2198	0.2203	0.2244	0.2299	-0.08
0 5 5 6 6 1	-0.7074	-0.7201	-0.7326	-0.7798	-0.8328	-0.418
0 5 6 5 6 1	-2.7451	-2.6786	-2.6151	-2.3879	-2.1485	-3.018
0 5 6 5 6 2	-1.5366	-1.4916	-1.4484	-1.2929	-1.1290	-1.575
0 5 6 6 6 1	0.5866	0.5541	0.5229	0.4101	0.2894	0.815
0 6 6 6 6 1	0.2867	0.3392	0.3907	0.5812	0.7923	-0.222
0 6 6 6 6 3	-2.2461	-2.1712	-2.1111	-1.8741	-1.6246	-2.435

In the last column of Table III we give the matrix elements calculated with the method of Brown and Kuo[1,2] and the Hamada-Johnston potential. In fact, we recently found a calculational error in the Brown and Kuo matrix elements[1-4], namely, the contribution from the $^3S_1$ second order long-range tensor force was overcounted by a factor of 2. But this overcounting was partly compensated because we underestimated the contributions from the $^3S_1$-$^3D_1$ cross terms. It is primarily this error which has initiated the present calculation. We feel that the present method is basically an improvement of our previous method[1-4]. But as shown by Table III, the present matrix elements are not qualitatively different from our previous matrix elements. Here we note that the appropriate value of ω for O^{18} and F^{18} is approximately -10 MeV.

IV.3 Energies of O^{18} and F^{18}.

In Figure 15, we list the various irreducible diagrams which we shall calculate for the vertex functions Γ, M and K. The function h of the same figure will be used to define the reaction-matrix Hartree-Fock spectrum for the single particle and hole lines, although h is in fact a part of the one-body vertex function Γ. Each wavy line in the figure represents a G-matrix, and hence each diagram of Figure 15 is a function of E, the same E as that of G(3412,E). Thus in the location of the poles of the Green's function, we need to solve the secular equation in a self-consistent way, as we discussed before. We have grouped the irreducible diagrams according to powers of (G-U). We note that up to $(G-U)^2$, there is no U for M and K. One main problem here is how will the vertex functions converge when we express them as a series of power of (G-U). With an appropriate choice of U, it may be possible to make the series converge. A systematic calculation of all the vertex-function diagrams up to $(G-U)^2$ is being carried out by S.Y. Lee and myself. This calculation is not yet finished.

In the present paper, I shall only report on some calculations which will illustrate the self-consistent procedure in the location of the poles of the Green's function. We have solved Eq. (20) with the following approximations. The one-body vertex functions are not calculated from the G-matrix elements. Instead, we have used the experimental[17] single-particle and single-hole spectrum. The vertex function M is calculated with the G-matrix elements calculated with the same method as that used for the calculation of the matrix elements of Table III. We first assume a trial energy E', then the ω variable for

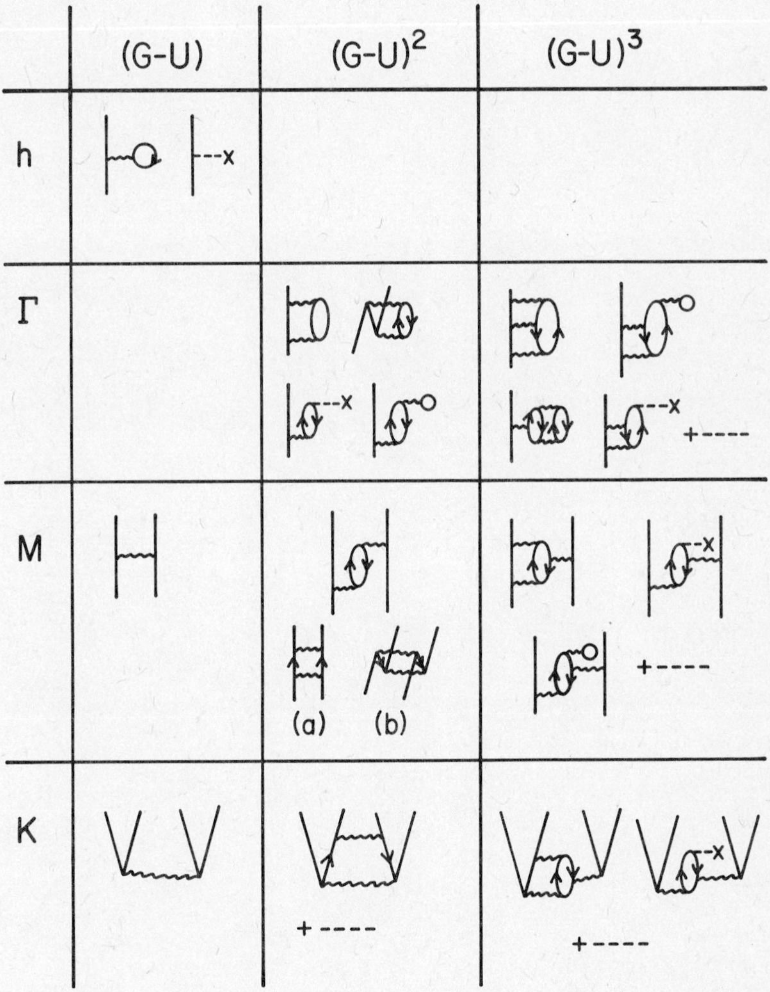

	(G-U)	$(G-U)^2$	$(G-U)^3$
h	--x		
Γ		--x	--x +----
M		(a) (b)	-x +----
K		+ ----	-x +----

Fig. 15. Irreducible diagrams which contribute to the vertex functions h,Γ, M and K. Each wavy line represents a G-matrix. Appropriate care for avoiding double counting must be taken when including the $(G-U)^2$ diagrams (a) and (b) in M.

TABLE IV

Energies of O^{18} and F^{18} calculated as a function of E'. In Case I, only the (G-U) term is included in M. In Case II, both the (G-U) and $(G-U)^2$ terms are included in M. The irreducible diagrams which are contained in M are shown in Figure 15. The results of Case II are in rather good agreement with experiment, and are very similar to the results reported in Ref. 2. All energies are in units of MeV.

			E'				
		J^π	-15	-13	-11	-9	-7
I.	O^{18}	0^+	-10.73	-10.74	-10.76	-10.77	-10.78
			- 8.37	- 8.38	- 8.38	- 8.39	- 8.39
		2^+	-10.11	-10.11	-10.12	-10.12	-10.13
			- 8.54	- 8.54	- 8.55	- 8.55	- 8.55
			- 4.15	- 4.15	- 4.16	- 4.16	- 4.16
II.	O^{18}	0^+	-12.04	-12.07	-12.09	-12.12	-12.15
			- 8.15	- 8.15	- 8.16	- 8.17	- 8.17
		2^+	-10.48	-10.49	-10.50	-10.51	-10.52
			- 8.40	- 8.40	- 8.40	- 8.40	- 8.40
			- 4.06	- 4.06	- 4.07	- 4.07	- 4.08
II.	F^{18}	3^+	- 8.37	- 8.41	- 8.45	- 8.49	- 8.53
			- 4.81	- 4.82	- 4.82	- 4.83	- 4.84

$G(\omega)$ is given by Eq. (46) with E replaced by E'. Then the secular equation is solved, with the eigenvalues obtained as a function of E'. Two calculations are made: (1) we include only the (G-U) term in M, namely, using the bare G-matrix only. (2) Both the (G-U) and the $(G-U)^2$ terms are included in M. Results are shown in Table IV. As shown, E varies very little as E' changes, within a reasonable range. The self-consistent solutions will be those E's with E=E'. They can be located easily by graphing E against E', and locating the points when E=E'. Thus we see that for the low-lying states of O^{18} and F^{18}, it is sufficient to use an E-independent G-matrix, namely the G-matrix elements can be calculated with <u>one</u> appropriate value of E'. But for some highly collective states such as the giant dipole states, it will probably be necessary to use an energy-dependent G-matrix and locate the energy levels self-consistently.

V. DISCUSSION

The method we described in the present paper is indeed an extension of the Brueckner-Goldstone theory for nuclear matter to finite nuclei. We have developed a well defined scheme for performing nuclear structure calculations starting from a realistic nucleon-nucleon potential and the non-relativistic Schroedinger equation. The following are some interesting problems which remain to be answered:

1. Convergence. It remains to be answered whether the energies given by the present method will converge or not as (a) we enlarge the model space in setting up the secular equation, and (b) we include higher and higher order irreducible diagrams in the vertex functions Γ, M and K. In the present method, correlations involving all particles are included to all orders as long as the particles are within the model space. But we only include the two-body correlations via the G-matrix when the particles are outside the model space. If we have chosen a large enough model space, that the nucleus is a low-density system and that the nuclear force is short-ranged should make the three and more-body correlations unimportant when the interacting particles are outside the model space. Further, when Γ, M and K are calculated according to powers of (G-U) there are two factors which may help in making the higher order diagrams small. First, the G-matrix elements are generally small compared with the energy denominators due to the particle-hole excitations of the core. Second, the choice of U is at our disposal, and we may choose U to suppress as much as possible the higher order terms. The above arguments are, however, at most qualitatively convincing. A quantitative study of the convergence problem remains to be done.

2. The calculation of the G-matrix. The present method of calculating the G-matrix by matrix inversion in the momentum space does appear to be very convenient and effective. It allows us to solve the G-matrix equation with a rather general projection operator in a straightforward way. We should, however, examine the NL-diagonal approximation which we have employed. The NL-non-diagonal contributions may be important for the density-dependence of the effective forces.

3. Energy-dependent or energy-independent effective interactions. If we refer the vertex functions in the present formalism as the effective interactions, then our effective interactions are energy dependent. Although we have found that this energy dependence is rather weak and smooth and hence can be handled rather conveniently in a self-consistent calculation of the nuclear energies, it would be highly desirable if we could devise an energy-independent effective interaction. In the folded-diagram formalism of Morita, Bandow and Baranger and Johnson[18], the effective interactions are energy independent but must be corrected by the folded diagrams. The folded-diagram formalism is designed mainly for perturbation calculation. It will be interesting to investigate whether the folded-diagram can be applied to the Green's function approach where the G-matrix is iterated to all orders within the model space. Barrett and Kirson[19] have calculated the G^3 contributions for the M vertex functions in the folded diagram formalism, and found that they are comparable in magnitude with the G and G^3 contributions. It will be interesting to investigate the relative importance of the (G-U), $(G\text{-}U)^2$ and $(G\text{-}U)^3$ contributions to the vertex functions in the present energy-dependent formalism[20].

4. The single-particle spectrum. Similar to the situation in the theory of nuclear matter, we can use the Bethe-Brandow-Petcheck theorem to define a G-matrix Hartree-Fock spectrum for the hole lines where the G-matrix is on the energy shell, namely independent of E and the other particle and hole lines which are present. But the situation for the particle lines is much less desirable. For the intermediate particle lines belonging to the G-matrix, we may use the free particle spectrum as we discussed earlier. But for the G-matrix Hartree-Fock spectrum for the particle lines within the model space, the G-matrix involved is off the energy shell, namely dependent on E and the other particle and hole lines which are present. This will introduce the requirement of the double self-consistency, namely both the vertex functions and the particle spectrum should be evaluated self-consistently. The requirement of the double self-consistency will cause considerable difficulty in calculation. In the calculation reported in this paper, this difficulty was avoided by using simply the experimental single-particle spectrum. It will be very desirable if we could devise a particle spectrum which depends only on the on-energy-shell G-matrix.

ACKNOWLEDGEMENTS

The work reported in this paper has been carried out in collaboration with K. Ratcliff and S.Y. Lee. It is my pleasure to thank them for their contributions and help in preparing the manuscript. I would like to thank Drs. G.E. Brown and A.D. Jackson for many stimulating discussions and criticisms and Sydel Blumberg for her valuable help in computation. It is my pleasure to acknowledge the hospitality of the Argonne National Laboratory where some of this work was carried out.

REFERENCES

1. T.T.S. Kuo, and G.E. Brown, Nucl. Phys. 85, 40 (1966).

2. T.T.S. Kuo, Nucl. Phys. A103, 71 (1967).

3. G.E. Brown and T.T.S. Kuo, Nucl. Phys. A92, 481 (1967).

4. T.T.S. Kuo and G.E. Brown, Nucl. Phys. A114, 241 (1968).

5. G.E. Brown, *Unified Theory of Nuclear Models*, North-Holland Publishing Co., Amsterdam, 1967.

6. M. Baranger, Proceedings of International School of Physics "Enrico Fermi", Vol. XL, p. 511, Academic Press, New York, 1967.

7. C.W. Wong, ibid, Vol. XL, p. 602.

8. M.H. Macfarlane, ibid., Vol. XL, p. 457.

9. Discussions on the Green's function method can be found in many text books and lecture notes on the theory of many-body problems. See, for example, B.R. Easlea, *Lectures on the Many-Body Problem*, (University of Pittsburgh, 1963, unpublished) and P. Nozieres, *The Theory of Interacting Fermi Systems*, W.A. Benjamin, Inc., New York, 1964.

10. See, for example, J.D. Bjorken and S.D. Drell, *Relativistic Quantum Fields*, (Chap. 17), McGraw-Hill Book Company, New York, 1965.

11. J. Goldstone, Proc. Roy. Soc. A239, p. 267, 1957. See also references 6 and 9 with regard to the cancellation of unlinked diagrams.

12. R.V. Reid, Annals of Physics 50, 411 (1968).

13. R.J. Eden and V.J. Emery, Proc. Roy. Soc. A248, 266 (1958).

14. H.A. Bethe, Phys. Rev. 138, B804 (1965); 158, 941 (1967).

15. *Handbook of Mathematical Functions*, ed. by M. Abramowitz and I.A. Stegun, Dover Publications, New York (1965).

16. G.E. Brown, A.D. Jackson and T.T.S. Kuo, Nucl. Phys. A133, 481 (1969).

17. See column (1) of Table I, T.T.S. Kuo, Phys. Letters 26B, 63 (1967).

18. M. Johnson and M. Baranger, preprint (MIT, 1970), and references quoted therein.

19. B. Barrett and M. Kirson, to be published.

20. This is being investigated by S.Y. Lee and T.T.S. Kuo.

DISCUSSION

KRIEGER: I believe that the conclusion of the Kirson and Barrett paper [Ref. 19] is that they were unable to find a way to order the diagrams such that the perturbation series demonstrated any convergence. Further, I believe that they purposely omitted the -U insertions, not for computational simplicity, but rather to compare with your published results, which do not include the insertion.

Although in the Green's function approach you do indeed sum a selected class of diagrams, this summation does not in itself answer the question raised by Barrett and Kirson. I think that it would still be necessary to show that you have found a meaningful way to order the diagrams which contribute, e.g. to the basic self energy insertion itself.

KUO: Well, I think that I cannot show the series will converge. I wish to point out that in the Barrett-Kirson calculation they do not include any diagrams with U in them. Which I think is not correct. From our preliminary results the U seems to have some cancelling effect, but we have not checked it out, and I choose not to present the results. One nice feature of the Green's function formalism is that I know how to calculate term by term systematically, but I really don't know if it will converge or not.

KRIEGER: But Bethe showed that you cannot re-order the perturbation theory in nuclear matter in terms of G, but finite nuclei is another problem.

KUO: We are in the process of investigating this.

HARVEY: Can I make the comment that there is another kind of self-consistency, namely, that the calculation of the effective interaction should be consistent with the calculation of every other operator of the system, e.g., the effective charge.

KUO: Thank you for pointing this out to us. In fact, we are investigating this problem.

HARVEY: Maybe Larry Zamick will be talking about this later.

HALBERT: Would you like to comment, Dr. Zamick?

ZAMICK: I would like to support the Barrett-Kirson calculation.

I did a casual matrix diagonalization where you take all the states which enter into the core re-normalization involving exciting a particle-hole. You can simply diagonalize the full matrix three particles, one hole and two particles, and when you do that you get a significant difference especially for J = 0 states than if you do the same calculation with perturbation theory, so I think maybe Kirson and Barrett are right.

HALBERT: Was this around calcium or where did you do this?

ZAMICK: Yes, two $f_{7/2}$ holes in Co^{54}. Then the particle-hole pair was limited to $(f_{7/2})^{-1}p_{3/2}$, $(f_{7/2})^{-1}f_{5/2}$, $(f_{7/2})^{-1}p_{1/2}$, that is only minor shell excitations.

KUO: I think I just want to make one more comment. In the Kirson-Barrett calculation they did not use an energy dependent G-matrix and I hope that the energy dependent G-matrix might make their diagram somewhat smaller. When you have some other particles and holes around, you will have, generally, a larger off-energy shell effect. That will make the matrix elements weaker.

HALBERT: But if there were a U that had a non-zero matrix element, you would have put it in?

KUO: Yes, as I discussed in my talk, all irreducible (G-U) vertex function diagrams will be included. Consider the following diagrams:

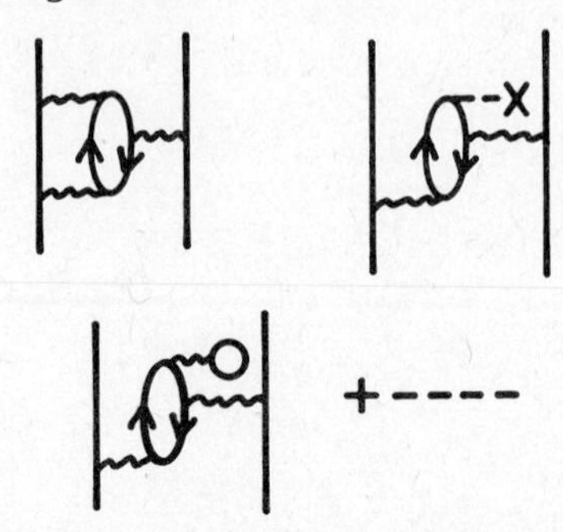

If the harmonic oscillator wave functions really satisfy the Hartree-Fock conditions then these two should cancel out. But since we are not using a Hartree-Fock self-consistent wave function, they will not cancel to zero. Both of these diagrams are not included in the Barrett-Kirson calculation.

ZAMICK: That's true.

KUO: To compute those we need a very complicated angular momentum recoupling. We have not yet finished our calculation.

ZAMICK: I think we can do an Oak Ridge-type matrix diagonalization of two particles and three particles-one hole and compare the two results.

KUO: Did you eliminate the spurious states?

ZAMICK: The problem of spurious states did not come up in my calculation.

HALBERT: Because of the particular shells you took into account--isn't that right?

ZAMICK: Yes. There was no excitation which changed the harmonic oscillator. This is for Co^{54} two $f_{7/2}$ holes, let the holes be in the f-shell and the particles in the rest of the f-p shell, $p_{3/2}$, $f_{5/2}$, and $p_{1/2}$. Then, if we do a matrix diagonalization, we get the J = 0+ state in Co^{54} depressed 500 keV less than in perturbation theory. Perturbation theory is equivalent to the Bertsch, Kuo-Brown core polarization.

KUO: I would like to interrupt. One more piece of information. The whole scheme is essentially that we are locating the poles but in a truncated fashion. There is no variational principle or other rule to tell us the relation between the energies obtained in our calculation and the energies of the true poles. We have calculated the binding energies of H^3 and He^4 in a $(OS)^3$ and $(OS)^4$ model space with the G-matrix method. The binding energies we found to be respectively 6 MeV and 20 MeV at $\hbar\omega$ = 14 and 19, respectively. When we vary the values of $\hbar\omega$, there are rather sharp minima around $\hbar\omega$ = 14 and $\hbar\omega$ = 19. These numbers look nice, but again I like to point out that I don't know whether they are the upper or the lower bounds for the true eigenvalues.

HALBERT: I have a question that I didn't get to ask. A long time ago you used to say that in calculating the bare-G matrix, you wanted specifically to use a Pauli operator which cut out perturbations from low-energy states just above the active shells, so that you could treat those perturbations in a better way, using Bertsch-type resonant matrix corrections, etc. Have you changed your mind?

KUO: No, I don't think I have changed; in the results I have presented I am using δ_{max} = 6 projection operator by which I mean that I am including all the f-p, s-d, 0_p, and 0_s shells in my model space. The approximation we made in the calculation of the G-matrix is probably safer if we use a larger model space.

LAWSON: What do you put in for the separation between the 2s-1d shell and 2p-1f shell?

KUO: In the present calculation, I just take the experimental single particle energy energies. But I will compute all the vertex functions and use them as my single particle energies.

II.B. LIFETIMES OF BOUND STATES*

R. E. Segel
Argonne National Laboratory, Argonne, Illinois
and
Northwestern University+, Evanston, Illinois

I. INTRODUCTION

There has been a great proliferation of lifetime data in the past couple of years, the reasons being, of course, the development of better and better lithium-drifted germanium detectors and the realization that the attenuated-Doppler-shift technique can be used to measure the lifetime of a large number of nuclear states. The time scale is of the order of 10^{-14} sec--just what is needed in order to measure the lifetimes of many bound states throughout the periodic table. There has also been a great improvement in gamma-ray spectroscopy measurements; in particular, there have been a good number of particle-gamma correlations from which mixing ratios have been extracted. There is so much data that a complete summary would be beyond the scope of this paper. Instead, the data are first lumped together and some overall features are presented, and then some individual nuclei are discussed--either because a crucial issue is involved or because the particular nucleus is representative of a number of cases that have been extensively studied.

II. AVERAGE TRANSITION RATES

A compilation of transition speeds of gamma rays emanating from low-lying bound states is given in Figure 1. Nuclei with $20 \leq A \leq 40$ are included, and "low-lying" is rather loosely and arbitrarily defined. Isospin-forbidden and isospin-unfavored transitions are excluded. It can be seen that all E1 transitions of this class are strongly inhibited, the average strength being around 10^{-4} of a single particle unit (as defined by Moszkowski[1]). This is a phenomenon observed throughout the periodic table. Virtually all of the E1 strength is up in the giant resonance, and therefore E1 decays of low-lying states are very strongly inhibited. There is an asymptotic selection rule[2] that in the limit of strong deformation E1 transitions are forbidden.

The M1 transition speeds show a large spread. A compilation such as Figure 1 is always biased, of course, in that

---

*Work performed under the auspices of the U. S. Atomic Energy Commission.

+Work supported in part by the National Science Foundation.

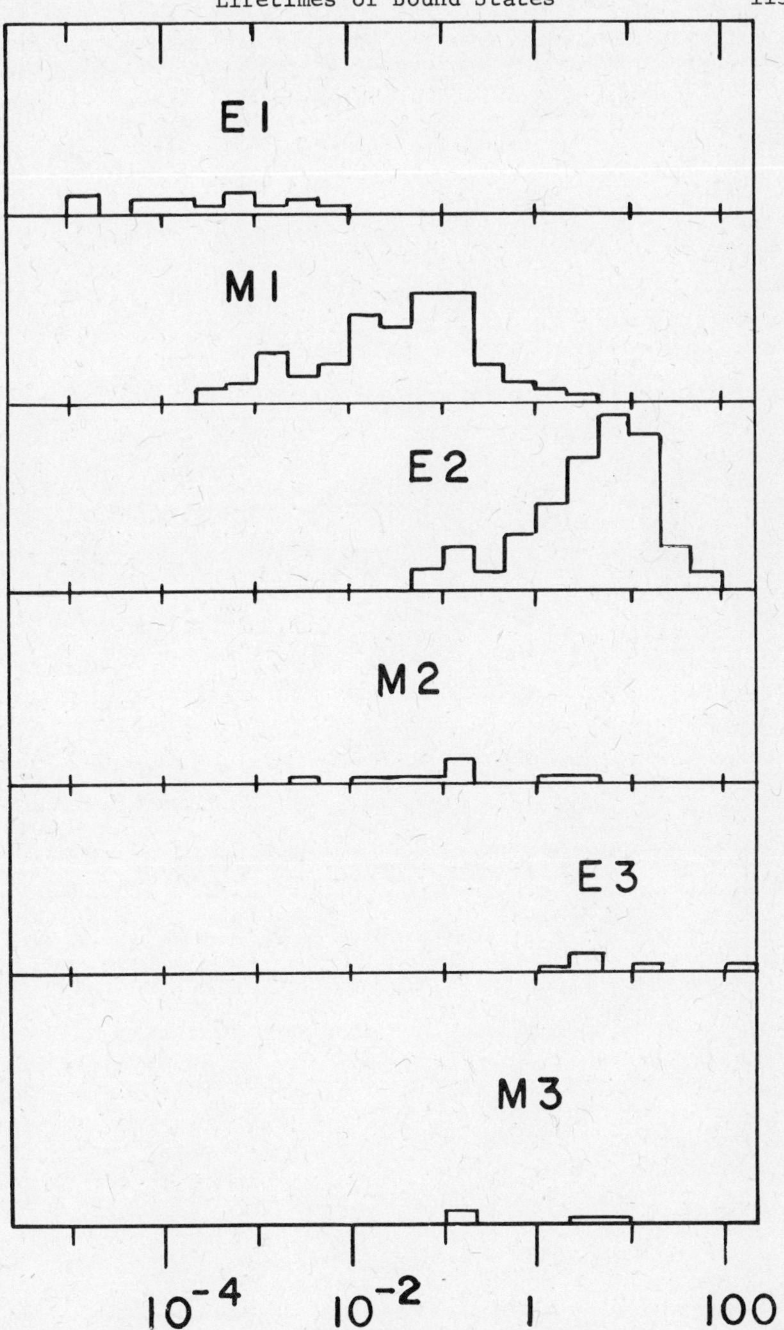

Fig. 1. Transition speeds, in Weisskopf units, of radiations between low-lying levels of nuclei with $20 \leq A \leq 40$. Isospin-forbidden and isospin-unfavored transitions have been omitted.

a lot of weak transitions are missed--particularly since transitions for which only an upper limit is known have not been included. The distribution peaks at about 10% of a single-particle unit and falls off very rapidly toward larger values; only a couple of the M1 transitions shown have speeds greater than 1 single-particle unit. On the other hand, many of the E2 transitions are enhanced, and the distribution peaks at an enhancement of about a factor of 10. (Again, many inhibited transitions have no doubt been missed). This E2 enhancement plays a very important role in the thinking about models of nuclei in the 2s-1d shell. The enhancements are attributed to collective motion, and their presence leads to the notion that rotational bands play a major role in the description of nuclei in this mass region.

The higher multipolarities are, of course, very much rarer, but there are now some eight M2 transitions whose speeds have been reported. Kurath and Lawson[3] have pointed out that the spatial properties of the M2 and the E1 operators are similar, and therefore the selection rule that holds for the E1 ought to hold for the M2 as well. They present impressive evidence that, over most of the periodic table, M2 transitions are indeed significantly inhibited. However, although many weak M2 transitions have no doubt been missed, the data in Figure 1 indicate that the rule is not very effective in the $20 \leq A \leq 40$ region where it is noteworthy that there are observed transitions whose reduced transition probabilities are very much greater than those for any of the E1 transitions in the region. Some of the fast transitions are understood; an example[3] is the transition in K^{39}. There is also room for experimental error because some of these transitions represent either weak branches or small admixtures. Nevertheless, from the evidence now available it must be said that, in contrast to the situation for E1 transitions, M2 transitions are not greatly inhibited for $A \leq 40$.

A few E3 transitions have now been measured, and they are all enhanced. This is as expected since, just as there is quadrupole enhancement, so too should there be octopole enhancement. Most of the reported cases are transitions from the first 3^- state to ground.

Four M3 transitions are included in Figure 1 and two of them are enhanced. These enhanced transitions are in Na^{24} and in Al^{24} and are probably accurately measured; the enhancement in Na^{24} is particularly well documented.[4] At first it might be thought that just as the M2 operator is similar to the E1, so too is the M3 similar to the E2; and since E2 transitions are often enhanced, enhancement might also be expected for M3 transitions. However this argument breaks down to some degree when the isospin properties of electromagnetic radiation are considered.

III. ISOSPIN EFFECTS

Neglecting higher order terms, the perturbing Hamiltonian can be written

$$H = \sum_{p_i} \frac{e}{c}\vec{v}_i \cdot \vec{A} + \sum_{n_i} \mu_n \vec{S}_i \cdot \nabla \times \vec{A} + \sum_{p_i} \mu_p \vec{S}_i \cdot \nabla \times \vec{A} \qquad (1)$$

where the sums are over the neutrons and protons in the nucleus and $\vec{A} \equiv \vec{A}(r_i)$ is the vector potential. By introducing isospin, Eq. (1) can be rewritten as a sum over all nucleons, namely

$$H = \sum_i \{ (\tfrac{1}{2} + \tau_{z_i}) (\frac{e}{c}\vec{v}_i \cdot \vec{A} + \mu_p \vec{S}_i \cdot \nabla \times \vec{A})$$

$$+ (\tfrac{1}{2} - \tau_{z_i}) \mu_n \vec{S}_i \cdot \nabla \times \vec{A} \} , \qquad (2)$$

where the particle-physics notation $\tau_z = +\frac{1}{2}$ for a proton and $\tau_z = -\frac{1}{2}$ for a neutron has been used. The perturbing Hamiltonian can then be written as the sum of two terms

$$H_0 = \tfrac{1}{2} \sum_i \frac{e}{c}\vec{v}_i \cdot \vec{A} + (\mu_p + \mu_n)\vec{S}_i \cdot \nabla \times \vec{A}, \qquad (3a)$$

$$H_1 = \sum_i \tau_{z_i} [\frac{e}{c}\vec{v}_i \cdot \vec{A} + (\mu_p - \mu_n)\vec{S}_i \cdot \nabla \times \vec{A}] . \qquad (3b)$$

H_0 is a scalar in isospin space and can therefore connect only states of the same isospin, while H_1 is an isovector and can contribute only when the vector triangle $\vec{T}_i + \vec{T}_f = \vec{1}$ can be closed.

Except for the obvious restriction $\Delta T < 2$, the only true isospin selection rule is for E1 transitions for which, as was first noted by Radicati,[5] in the long-wavelength limit the leading term in H_0 reduces to a sum over the velocities of all of the nucleons; and this sum will not contribute as it represents the center-of-mass motion. Thus, E1 radiation behaves as a pure isovector and therefore transitions between two isospin-zero states are forbidden. There is considerable evidence that this rule is effective in the sd shell. Isospin-forbidden transitions are, on the average, slower than isospin-allowed E1 transitions by a factor of 10 or perhaps somewhat more. Quantitatively, it is rather hard to interpret this inhibition. It certainly does not mean that isospin is good only to a factor of 10 in the sd shell. The problem is that the E1 transitions are already so inhibited that it is hard to see what the effect of any additional isospin inhibition would be; when two inhibitions are operating, they do not necessarily operate in tandem.

In the general case of a transition between two states of the same (nonzero) isospin, both the isoscalar and the isovector parts will be present. In the special cases in which one of them is absent, as is the case for E1 transitions, the reduced transition rates for mirror transitions should be the same. There is at present little data, but the comparison of mirror E1 transitions might be a good place to look for isospin violation in the sd shell. Observable effects might be present since on the one hand, the Coulomb forces in the sd shell are large enough to be significant, while on the other hand, the large neutron excess that often preserves isospin in nuclear reactions is not present.

In order to discuss isospin effects in magnetic transitions it is necessary to write down the specific operators that induce transitions of the various multipolarities. For M1 transitions,

$$H_0 = \tfrac{1}{2} \sum_i [\vec{\ell}_i + (\mu_p + \mu_n)\vec{S}_i] \tag{4a}$$

$$= \tfrac{1}{2}(\mu_p + \mu_n - \tfrac{1}{2}) \sum_i \vec{S}_i = 0.19 \sum_i \vec{S}_i, \tag{4b}$$

where (4b) was deduced from (4a) by Morpurgo,[6] who concluded that the small multiplying factor in (4b) leads to an inhibition of isoscalar M1 transitions. In contrast, the isovector part can be written

$$H_1 = \sum_i \tau_{z_i} [\vec{\ell}_i + (\mu_p - \mu_n) \vec{S}_i] = \sum_i \tau_{z_i} (\vec{\ell}_i + 4.70 S_i) , \tag{5}$$

and therefore on the average it can be expected that isoscalar M1 transitions will be retarded by about a factor of 100 relative to isovector M1 transitions. There is considerable evidence that Morpurgo's rule is obeyed and that isoscalar M1 transitions are inhibited in the sd shell. In view of the large spread in the rates of allowed M1 transitions, the degree of isospin conservation is difficult to assess quantitatively on the basis of the observed transition rates.

For mirror transitions, we can write

$$\frac{\Gamma_{\tau_z}}{\Gamma_{-\tau_z}} = \left(\frac{H_0 + H_1}{H_0 - H_1}\right)^2 . \tag{6}$$

Therefore even if the isoscalar intensity is small compared to the isovector, the ratio of mirror transition rates can differ substantially from unity; if H_0 is 10% of H_1, the transition rates will differ by 50%. Thus, we can learn little about isospin by comparing mirror M1 transitions. What can be done, however, is to use Eq. (6) to extract the isovector and isoscalar amplitudes from the measured rates.

It has been pointed out[7] that the relatively large factor multiplying the spin part of the isovector amplitude makes possible a comparison between the Gamow-Teller beta decay, which is a pure spin flip, and the M1 decay of the analog state. Again, this is a rather crude test of isospin conservation. Rather, it is more profitable to assume isospin conservation and to use the comparison to determine the contribution from the orbital part. A number of cases have now been studied, and in some of these the orbital contribution is great enough to raise the M1 transition rate to several times what can be predicted from the beta decay.[8]

It has been shown by Sugimoto[9] that the above line of reasoning can be carried further by combining the beta-decay rate between mirror states and their magnetic dipole moments. It is assumed that the states have good isospin. Four related quantities can be experimentally determined: the two magnetic moments, which will in general be different, the spin of the two states, and the log ft of the beta decay. Therefore, from the four experimental numbers the four amplitudes (the isovector and isoscalar components of the orbital and spin parts) can be determined. The beta decay will have both Fermi and Gamow-Teller contributions; but since the transitions are between analog states, the overlap integral for the Fermi part is unity and hence the Gamow-Teller matrix element can be extracted. Sugimoto has published an analysis of the sd-shell data.[9]

Warburton[10] has pointed out that the isoscalar parts of all magnetic transitions are inhibited by a factor similar to that inhibiting isoscalar magnetic-dipole transitions. The subject of the role of isospin in electromagnetic transitions has been reviewed recently by Warburton and Weneser.[11] Table I gives the ratio of the coefficients multiplying the isoscalar and isovector parts of magnetic transitions. It can be seen that the inhibition gets only slightly relaxed as one goes to higher multipolarities. This is really the source of the trouble in trying to understand fast M3 transitions; if it is expected that there is an enhancement similar to the E2 enhancement, it must be remembered that the E2 enhancement is believed to be present when transitions can be described as between members of the same rotational band so that the enhancement is restricted to the isoscalar part. Thus while collective enhancement of M3 transitions is possible, the part that is enhanced (namely the isoscalar) has a single-particle transition rate that is only about 1/70 of a Weisskopf unit.

TABLE I.
Average ratio of isoscalar to isovector amplitude for ML transitions

L	H_0/H_1
1	0.091
2	0.122
3	0.138

Figure 2 shows the fast M3 transitions. The lifetimes have been accurately determined,[4] particularly the state in Na^{24}, and it is very unlikely that the apparent enhancement is attributable to experimental error. If the Na^{24} transition were pure isoscalar, a collective enhancement of about a factor of 400 would be required. However, the very fact that these strong mirror transitions have substantially different reduced rates means that both the isoscalar and the isovector contributions must be significant. There appears to be significant enhancement, and it would be nice to know where it is coming from. It should be remarked that the definition of a single-particle unit is somewhat arbitrary; before it is really established that there is something wrong here, it may be necessary to more precisely define just what is to be expected for a single-particle transition in these cases. The statistical factors must be put in correctly, and this is not a completely trivial task, and also the correct combinations of magnetic moments must be used. Perhaps then it will be possible to reproduce the observed transition rates without introducing any unreasonable enhancements.

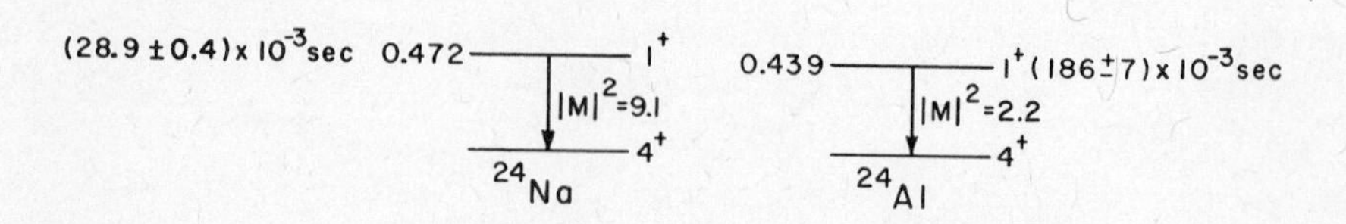

Fig. 2. M3 transitions in Na^{24} and Al^{24}.

IV. ROTATIONAL REGION

We now pass to the type of investigation that has received most of the effort in this area in the past few years, namely collecting extensive data in various nuclei and attempting to fit the data with various models. This works best in the A ≈ 25 region of the sd shell, where the rotational model seems to work very well. The mass-25 nuclei themselves have been discussed previously in some detail.[12] The rotational bands are based upon Nilsson orbitals, the relevant ones of which are shown in Figure 3. One of the reasons why Mg^{25} and Al^{25} are such favorable cases is that the $\frac{1}{2}^+$ and $\frac{3}{2}^+$ orbits are filled and, since the deformation is prolate, the odd neutron or proton is in a $\frac{5}{2}^+$ orbit and the next higher state for it is $\frac{1}{2}^+$. While these latter two states are rather close together in energy, the spin difference is great enough so that there is little band mixing; there is a nice clean $K = \frac{5}{2}^+$ ground-state band in Mg^{25} and Al^{25}, also a low-lying $K = \frac{1}{2}^+$ band. The E2 transitions within the band are enhanced by about a factor of 10. In contrast, interband E2 transitions have somewhat less than single-particle speed, usually a few tenths of single-particle speed.

Because intraband transitions are between states of the same basic configuration, the rates can be expressed in terms of static moments and the relative rates depend only on Clebsch-Gordan coefficients. For E2 transitions,

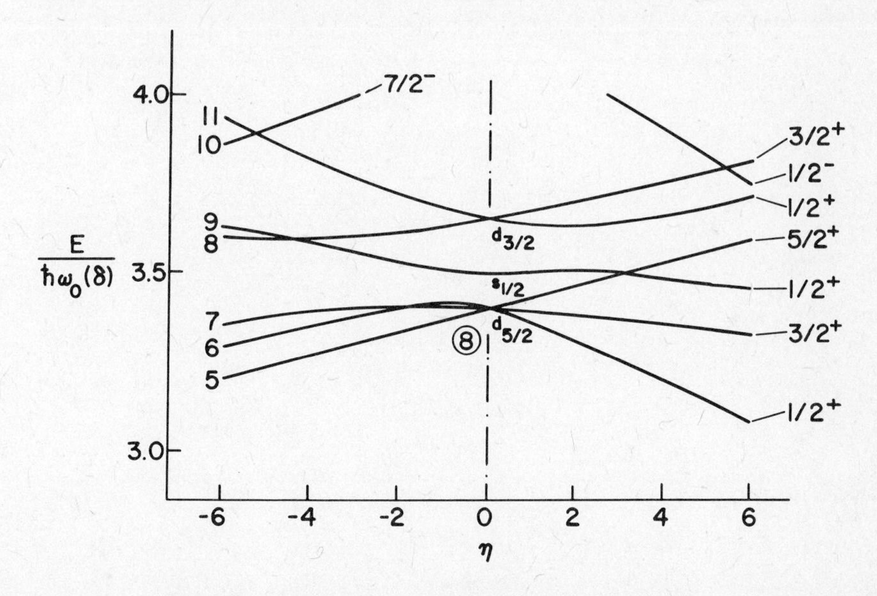

Fig. 3. Nilsson orbits in the sd shell.

$$B(E2) = \frac{5}{16\pi} (J_i 2K0|J_f K)^2 e^2 Q_0{}^2, \quad (7)$$

where Q_0 is the quadrupole moment. Equation (7), rewritten in terms of single-particle units, becomes

$$|M(E2)|^2 = (3.472 \times 10^4/(1.2)^4 A^{4/3})(J_i 2K0|J_f K)^2 Q_0{}^2, \quad (8)$$

with Q_0 in units of 10^{-24} cm^2. Similarly, for intraband M1

$$B(M1) = \frac{3}{4\pi} K^2 (J_i 1K0|J_f K) \mu_0{}^2 (g_K - g_R)^2, \quad (9)$$

where g_K and g_R are the g-factors for the core and the valence nucleons, respectively. In single-particle units this becomes

$$|M(M1)|^2 = \frac{2}{15} K^2 (J_i 1K0|J_f K)(g_K - g_R)^2. \quad (10)$$

A good quantitative test of the rotational-band picture is that the same quadrupole moment must be extracted from all E2 transitions within a given band and that the same $(g_K - g_R)$ be extracted from the M1 transitions.

A number of nuclei between about mass 19 and mass 26 have been shown to contain well defined rotational bands. A good case in point is Na^{22}. The three lowest members of the $K = 3^+$ Na^{22} ground-state band are shown in Figure 4. The lifetimes, branching ratios, and mixing ratios (including the sign) have all been measured for these states.[13] Thus the M1 and E2 transition amplitudes can be determined. As can be seen from Table II, the same Q_0 (within experimental error) was determined from all three E2 transitions. The M1 is an isospin-unfavored transition and is inhibited, and yet

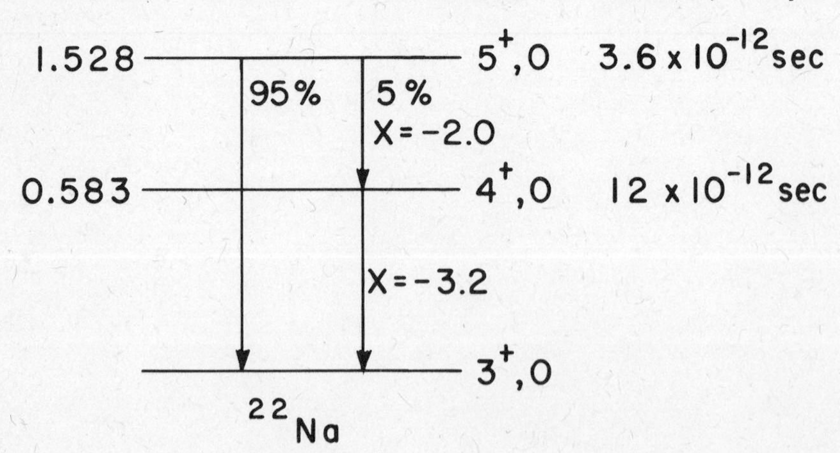

Fig. 4. The $K = 3^+$ ground-state band in Na^{22}. The data are from Ref. 13.

TABLE II

Transition rates in the $K = 3^+$ band in Na^{22}[a].

J_i^π	J_f^π	$\lvert M(M1)\rvert^2$	$(g_K - g_R)$	$\lvert M(E2)\rvert^2$	Q_0[b]
5^+	4^+	$(3.0 \pm 1.1) \times 10^{-4}$	$+0.025 \pm 0.004$	20.7 ± 7.0	0.48 ± 0.08
5^+	3^+			$6.9^{+2.0}_{-1.2}$	0.55 ± 0.06
4^+	3^+	$(3.3 \pm 0.6) \times 10^{-4}$	$+0.029 \pm 0.003$	29.5 ± 5.0	0.56 ± 0.05

[a]Reference 13.

[b]Assumed positive.

FROM MAGNETIC MOMENT MEASUREMENTS:

$$g_K - g_R = [(0.596 \pm 0.003) - (0.540 \pm 0.009)] = 0.056 \pm 0.013$$

the two transitions agree well and the resulting value of $(g_K - g_R)$ is in satisfactory agreement with the measured moments. The sign of the mixing ratio is correctly predicted by the rule

$$\text{sgn } x = - \text{ sgn } \frac{(g_K - g_R)}{Q_0} \tag{11}$$

if the quadrupole deformation is taken to be positive, as is the case wherever it has been measured in nuclei with atomic number near 22.

In other cases the M1 transitions are less inhibited. Table III lists data from Ne^{21}, for which the first five members of the ground-state band have been identified.[14] Here the M1 rates are of the order of 10% of single-particle speed, which is near the upper end of the M1 rate distribution. The relative transition rates are in satisfactory agreement with Eqs. (7)--(10), although the agreement is just about on the edge of experimental error. Scattered throughout the mass region $19 \leq A \leq 26$ are a number of cases for which the rotational picture gives good quantitative predictions for relative transition probabilities. Unfortunately, there is at least one case that does not seem to work out very well, and that is Na^{23}.

Figure 5 gives the energy-level diagram for Na^{23}. The $\frac{3}{2}^+$ ground state and the $\frac{5}{2}^+$, $\frac{7}{2}^+$, and $\frac{9}{2}^+$ excited states appear as if there is well defined $K = \frac{3}{2}^+$ band. However, the relative transition rates do not fit the picture.

Transition rates reported[15,16] in Na^{23} are shown in Figure 6. It can be seen that while the $\frac{5}{2}^+ \rightarrow \frac{3}{2}^+$, $\frac{9}{2}^+ \rightarrow \frac{5}{2}^+$, and $\frac{9}{2}^+ \rightarrow \frac{7}{2}^+$ E2 transitions are all strongly enhanced, the $\frac{7}{2}^+ \rightarrow \frac{3}{2}^+$ and $\frac{7}{2}^+ \rightarrow \frac{5}{2}^+$ transitions are much weaker. After the Clebsch-Gordan coefficients have been put in, the relative transition rates disagree by at least a factor of five. The M1 transitions also show a discrepancy in that the $\frac{7}{2}^+ \rightarrow \frac{5}{2}^+$ is significantly slower than the other two. Perhaps the trouble is due to experimental error; it is suspicious that all three transitions from the $\frac{7}{2}^+$ state are much slower than the other intraband transitions. Some discrepancies have appeared in the lifetimes reported[16] for other states in Na^{23}. Only one group[15] has reported a value for the 440-keV state, and it therefore would be nice if it could be remeasured.

Even allowing for experimental error, the region in which the simple rotation picture is valid terminates at around mass 27. Figure 7 shows the transition rates[4] in Al^{27}. Just from looking at the energy-level diagram, it appears probable that some sort of rotational-band pattern would fit the spectrum; but it is clear that the E2 transition rates would not be reproduced. In particular, all of the observed E2 transitions are enhanced. Of course the observed E2 rates can be taken as evidence of strong band mixing, since it is always possible to reproduce the experimental data if enough

TABLE III

Transition rates in the $K = \frac{3}{2}^+$ band in Ne^{21}[a].

J_i^π	J_f^π	$\lvert M(M1)\rvert^2$	$\lvert g_K - g_R\rvert$	$\lvert M(E2)\rvert^2$	Q_0
$\frac{11}{2}^+$	$\frac{9}{2}^+$	0.07 ± 0.03	0.77 ± 0.14	3.9 ± 2.7	0.37 ± 0.14
$\frac{11}{2}^+$	$\frac{7}{2}^+$			12.6 ± 4.3	0.41 ± 0.08
$\frac{9}{2}^+$	$\frac{7}{2}^+$	0.12 ± 0.03	1.01 ± 0.10	5.9 ± 2.7	0.38 ± 0.10
$\frac{9}{2}^+$	$\frac{5}{2}^+$			6.3 ± 1.6	0.32 ± 0.06
$\frac{7}{2}^+$	$\frac{5}{2}^+$	0.07 ± 0.02	0.83 ± 0.12	7.1 ± 2.8	0.34 ± 0.08
$\frac{7}{2}^+$	$\frac{3}{2}^+$			4.8 ± 1.2	0.34 ± 0.06
$\frac{5}{2}^+$	$\frac{3}{2}^+$	0.03 ± 0.01	0.65 ± 0.10	5.0 ± 2.8	0.27 ± 0.09

[a]Reference 14.

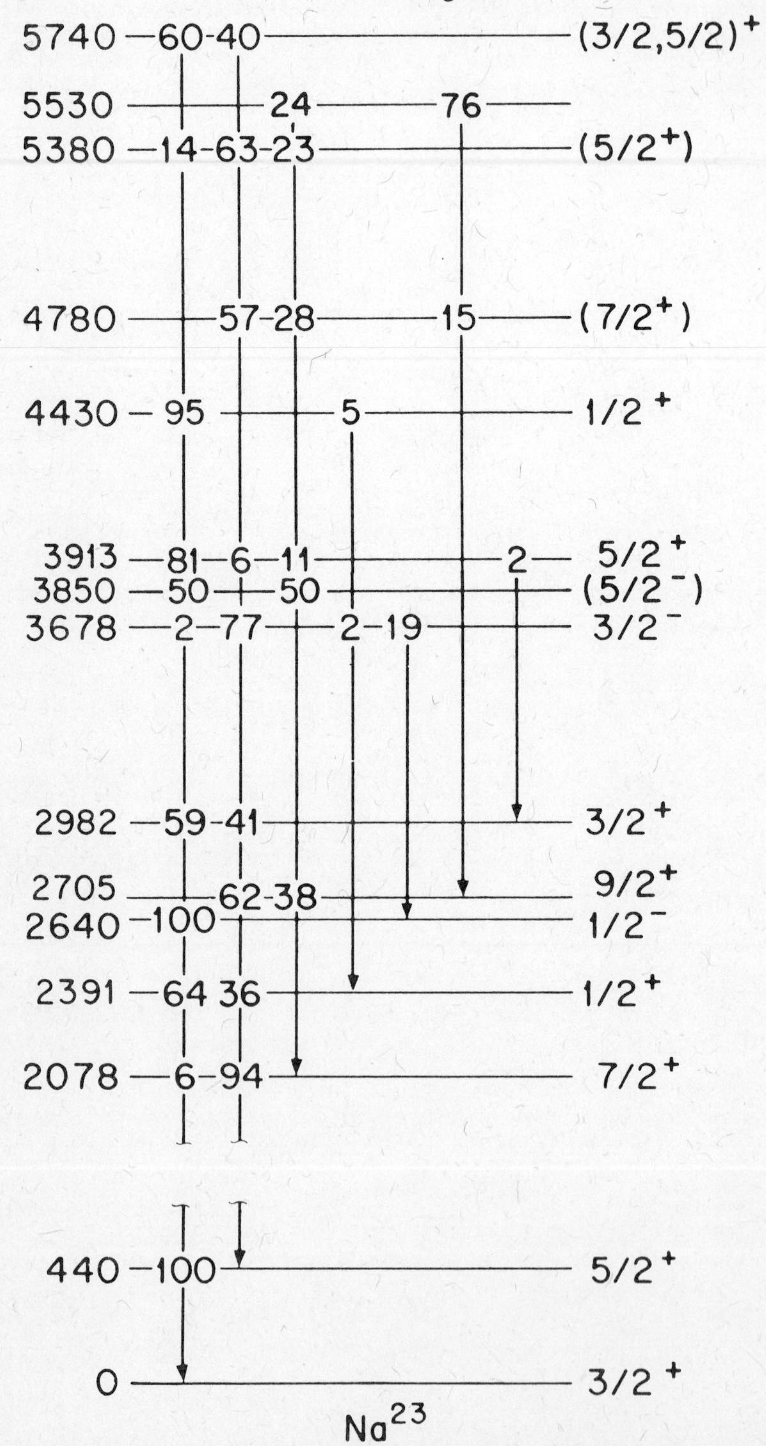

Fig. 5. Energy level diagram of Na^{23}.

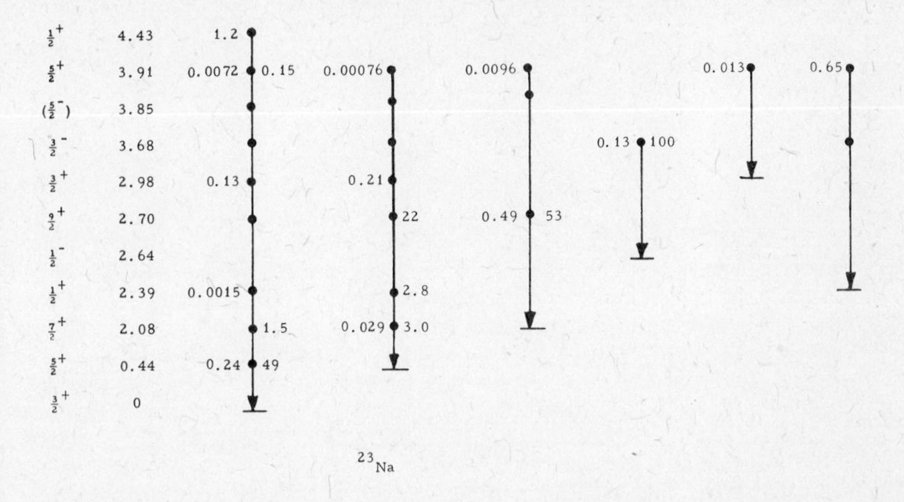

Fig. 6. Transition rates in Na^{23}, expressed in single-particle units (Ref. 1). The M1 rate is given on the left of the line and the E2 rate on the right. If no rate is given, either it is not known or the transition does not occur. The data are from Ref. 15.

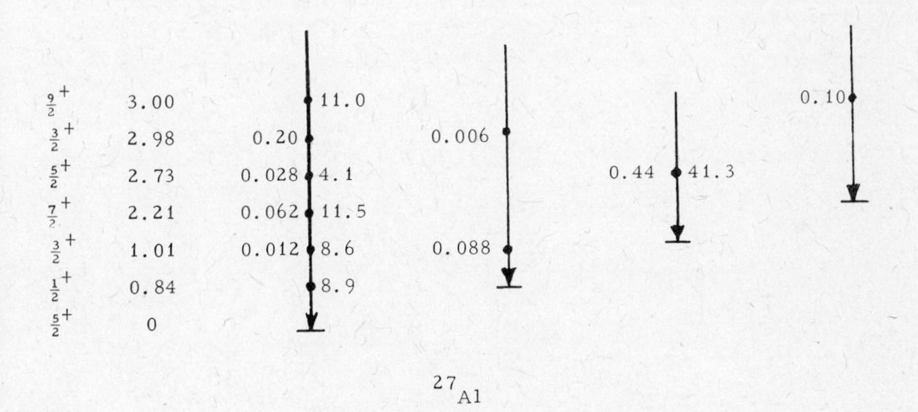

Fig. 7. Transition rates in Al^{27}.

bands are mixed in. However, the calculations then become too unwieldy--and while the model does not lose its validity, it does lose its utility.

V. SPHERICAL REGION: EFFECTIVE CHARGES AND EFFECTIVE MOMENTS

Even if the rigid-rotator model could be extended further into the shell, it must be remembered that the Nilsson model is an extension of the simple shell model. At the ends of the shell it is expected that the spherical shell model is valid. Thus what is really desired is to be able to theoretically understand the transition rates in the nuclei near the end of the shell where the shell-model configurations are simple and then allow for the collective motion in the more complex nuclei. However if we try to carry out this program, we immediately run into trouble, as can be seen by looking at the data in the "closed shell plus one" nuclei O^{17} and F^{17} (Figure 8). The spectroscopy looks fine in that the spins and parities of the two lowest states are just what is expected from the spherical shell model and there is then a substantial energy gap below the second excited state. How-

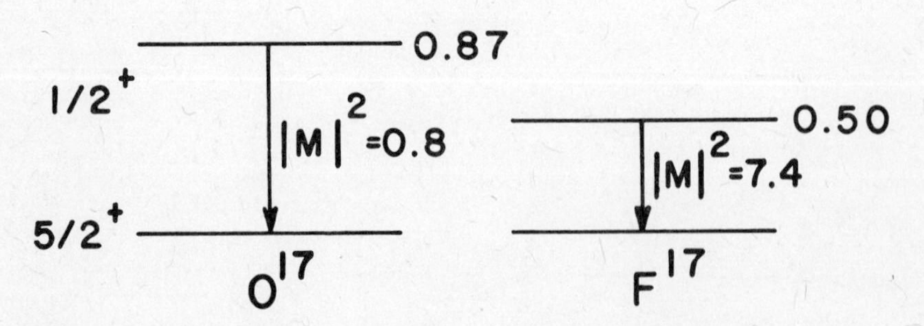

Fig. 8. E2 transitions in O^{17} and F^{17}.

ever the E2 transition rates from the first excited state to ground are obviously too large to fit this simple picture.[17] In fact, except for a tiny part due to the requirement that there be no center-of-mass motion, if O^{17} can be described as a closed shell plus an odd neutron there should be no electric transition at all. Similarly the $\frac{1}{2}^+ \rightarrow \frac{5}{2}^+$ rate in F^{17} is too rapid to be due to a single proton and some enhancement is required in order to explain the observed rate. Such enhancements are seen throughout the shell, and what is usually done in order to explain the observed E2 rates is to endow the neutron with an effective charge and to add an effective charge to the proton. This procedure is justified by invoking a picture in which some motion in the core is correlated with the motion of the valence nucleon. The effective charge that is added to the real charges is usually between one half and one times the proton charge.

The situation is similar at the other end of the shell (Figure 9). Again, the spins of the ground state and first excited state are those expected for an odd nucleon orbiting a closed-shell Ca^{40} core. The situation is somewhat more complicated here because the $\frac{3}{2}^-$ state seems to be split, presumably because a $\frac{3}{2}^-$ state at about the same energy can also be formed by adding an $f_{7/2}$ nucleon to the first 2^+ state in Ca^{40}. In each of these mass-41 nuclei, then, the two low-lying E2 $\frac{3}{2}^- \rightarrow \frac{7}{2}^-$ transitions contain two major contributions: one from the odd nucleon and the other from the $2^+ \rightarrow 0^+$ core transition. It therefore appears[18] that the two amplitudes interfere constructively in the transition from the lower $\frac{3}{2}^-$ state but destructively in that from the upper. In fact, the cancellation is almost complete and as a result the upper $\frac{3}{2}^- \rightarrow \frac{7}{2}^-$ transitions are about the slowest E2's which are not violating a known selection rule. This picture implies that the single-particle $\frac{3}{2}^- \rightarrow \frac{7}{2}^-$ amplitude is similar to that of the $2^+ \rightarrow 0^+$ core transition; the required enhancement can be reproduced if the nucleons are

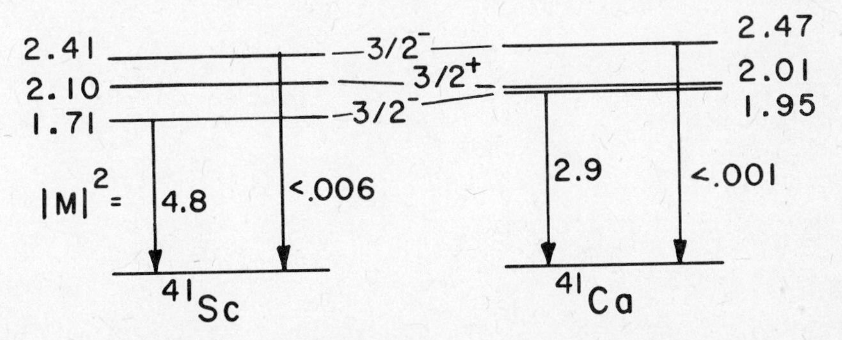

Fig. 9. Low-lying levels in Ca^{41} and Sc^{41}.

endowed with an effective charge. The data are fitted rather well by calculations by Gerace and Green,[19] in which an effective charge is added to each nucleon.

Just as it has been found desirable to use effective charges in calculating electric transitions, effective moments have been used in calculating magnetic transitions. At the upper end of the shell there is a very good opportunity to quantitatively assess how accurately transition rates can be reproduced by using effective moments. The spectrum of K^{40} is shown in Figure 10. The simple shell-model picture would describe K^{40} as $(\nu f_{7/2})(\pi d_{3/2})^{-1}$, and indeed the spectrum looks right in that the four lowest states have the spins and parities expected from this configuration. At higher energies there are positive-parity states and other negative-parity states, and the configurations that have been assigned[20-23] to the various states are as shown in Figure 10. Of course, qualitatively reproducing the level scheme is not good evidence that a model is valid, but in this case there is quantitative evidence that these states are well described as $(\nu f_{7/2})(\pi d_{3/2})^{-1}$.

In particular, it was shown by Pandya[20] and independently by Talmi and Goldstein[21] that if K^{40} is taken to be $(\nu f_{7/2})(\pi d_{3/2})^{-1}$ and Cl^{38} is taken to be $(\nu f_{7/2})(\pi d_{3/2})$, then the spectra of these two nuclei are simply related.

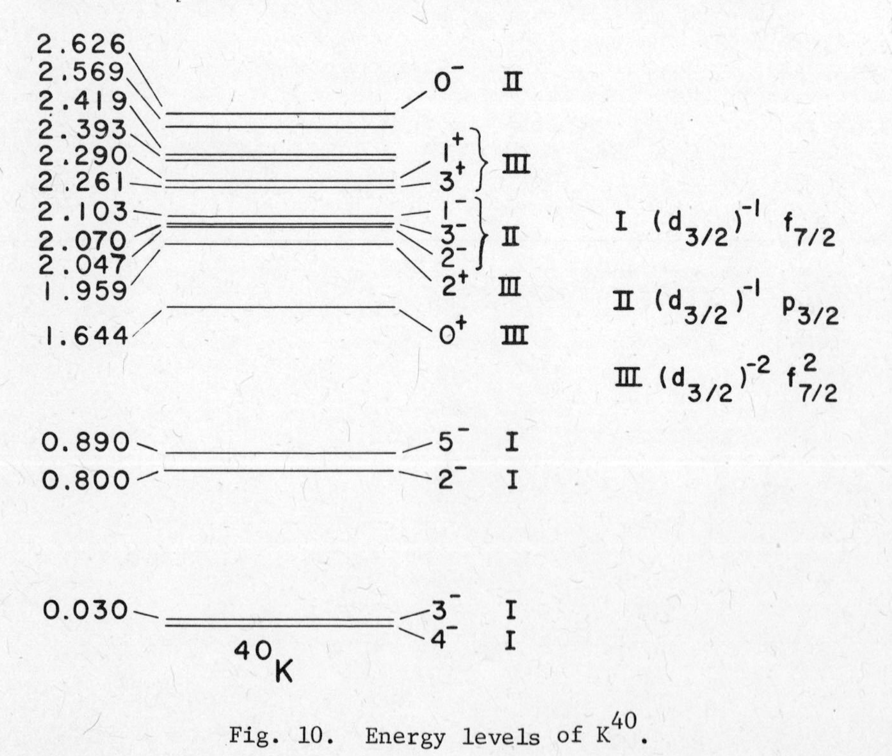

Fig. 10. Energy levels of K^{40}.

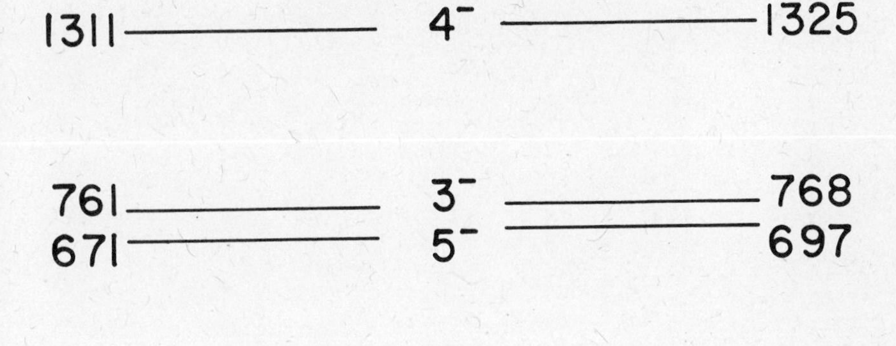

Fig. 11. Comparison of the transformed K^{40} levels, denoted $P(K^{40})$ with the observed Cl^{38} levels. The transformation is according to Eq. (12). The lowest transformed K^{40}state has been set at zero energy.

These authors showed that a particle-particle spectrum can be derived from a particle-hole spectrum (or vice versa) by the relationship

$$E_{J'} = \sum_J (2J + 1)\ W\ (j_1 j_2 j_2 j_1; J'J) E_J . \qquad (12)$$

In Figure 11 the observed Cl^{38} level scheme is compared with that derived from K^{40} by use of the Pandya transformation, Eq. (12). The excellent agreement has been taken as evidence that the low-lying states of Cl^{38} and K^{40} are well described as being members of the $(\nu f_{7/2})(\pi d_{3/2})$ quartet.

The speed of an M1 transition between two members of the quartet is given by the formula

$$\frac{1}{\tau} = (1.673 \times 10^{13}) E_\gamma{}^3 (2J_f + 1)[(-1)^{J_i}\sqrt{2} W(J_i J_f \tfrac{7}{2}\tfrac{7}{2}; 1\tfrac{3}{2})$$

$$\times <\nu f_{7/2}||\vec{\mu}||\nu f_{7/2}> + (-1)^{J_f} W(J_i J_f \tfrac{3}{2}\tfrac{3}{2}; 1\tfrac{7}{2}) <\pi d_{3/2}||\vec{\mu}||\pi d_{3/2}>]^2 \qquad (13a)$$

$$= (3.345 \times 10^{13}) E_\gamma (2J_f + 1) W^2 (J_i J_f \tfrac{7}{2}\tfrac{7}{2}; 1\tfrac{3}{2}) [(\nu f_{7/2}||\vec{\mu}||\nu f_{7/2}>$$

$$- (21/5)^{1/2} <\pi d_{3/2}||\vec{\mu}||\pi d_{3/2} J]^2 \qquad (13b)$$

$$= (3.345 \times 10^{13}) E_\gamma{}^3 (2J_f + 1) W^2(J_i J_f \tfrac{7}{2}\tfrac{7}{2}; 1\tfrac{3}{2}) \; |M_{df}|^2, \quad (13c)$$

where E_γ is in MeV and τ is in seconds. Equation (13b) is obtained from (13a) by using the relationship

$$\frac{W(J_1 J_2 J_a J_a; 1 J_b)}{W(J_1 J_2 J_b J_b; 1 J_a)} = (-1)^{2(J_a - J_b)} \left[\frac{J_b(J_b + 1)(2J_b + 1)}{J_a(J_a + 1)(2J_a + 1)}\right]^{1/2} \quad (14)$$

with $|J_1 - J_2| = 1$. It is apparent from Eq. (13b) that the model fixes the relative transition rates within the multiplet independent of what moments are ascribed to the valence nucleons; i.e., the same $|M_{df}|^2$, defined by (13c), should be obtained from all three M1 transitions.

Lifetimes in K^{40} and Cl^{38} have been measured at Argonne.[24] Figure 12 shows the experimental arrangement. The nuclei are made by the (d,p) reaction, and coincidence spectra are taken between a fixed gamma-ray detector and proton detectors. One proton detector is at 0°, where it receives protons from reactions in which very little momentum is given to the recoil nucleus and therefore the unshifted line is in coincidence. The other is set so that the recoil is moving along the axis of the gamma-ray detector.

Data from K^{40} are shown in Figure 13. The resolution width is about 5 keV, and both of the gamma rays of Figure 13 would show a full shift of about 4.5 keV. For each state the shift is attenuated to about 20-30% of its full value when the recoils are slowed down in KI. The lifetime measurements are summarized in Table IV. The expected shift attenuation as a function of the lifetime of the state was calculated for each state by use of the standard theory of the slowing down of ions in matter. When it became apparent that the K^{40} results were not coming out in agreement with Eq. (13), a number of checks were made to ensure that the discrepancy was not due to experimental error. The degree of attenuation for the two states in K^{40} was determined several times. In addition there was good agreement in the lifetimes as measured with two different stopping materials whose densities differed by a factor of 5.

The data in Cl^{38}, which are more complete than those given in Ref. 24, also do not yield lifetime ratios (Table V) in agreement with Eq. (13). In Cl^{38} the branching ratio from the 1.311-MeV state provides an independent check on the validity of Eq. (13). A measurement of this branching ratio is shown in Figure 14. From Eq. (13) the ratio of the intensity of the 640-keV gamma ray to that of the 550-keV gamma ray is predicted to be 1.2. In contrast, after correcting for the variation of detector efficiency with energy, the spectrum of Figure 14 yields the ratio 3.7 ± 0.5. These data were taken in coincidence with a proton detector that was placed along the beam direction so that the beam direction is the only axis of symmetry. Furthermore, the gamma-ray detector was placed at 125°, at which angle the second Legendre polynomial vanishes. Since E2 enhancements of at least a hundred would have to be present in order for the P_4 term to be significant, the measurement shown in Figure 14 is a valid measure of the branching ratio. Again, since the E2 intensity is almost certainly small compared to that of the M1, the measured branching ratio can be taken to be the M1 branching ratio.

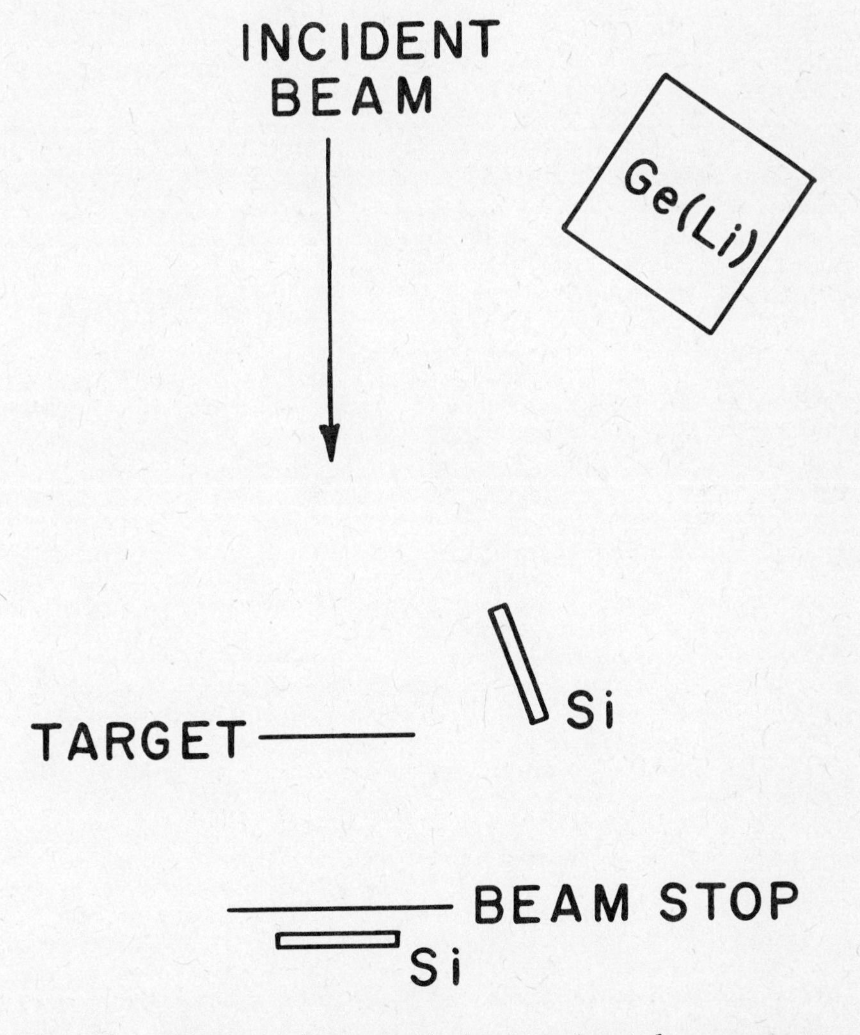

Fig. 12. Schematic diagram of the experimental arrangement for making attenuated-Doppler-shift measurements.

TABLE IV

Mean lives of states that are considered to be members of a $(\pi d_{3/2})(\nu f_{7/2})$ quartet. The quality F is the ratio of the measured shift to the unattenuated shift.

Nucleus	E_x (MeV)	E_γ (MeV)	Stopping material	F	Measured τ (psec)	Adopted τ (psec)
^{40}K	0.800	0.770	KI	0.36 ± 0.04	0.70 ± 0.15	0.65 ± 0.15
^{40}K	0.800	0.770	Au	0.15 ± 0.06	0.60 ± 0.25	
^{40}K	0.890	0.890	KI	0.18 ± 0.03	1.6 ± 0.3	1.6 ± 0.3
^{40}K	0.890	0.890	Au	0.00 ± 0.06	> 1.4	
^{38}Cl	0.761	0.761	$PbCl_2$	0.40 ± 0.03	0.37 ± 0.07	0.33 ± 0.07
^{38}Cl	0.761	9.761	Au	0.28 ± 0.05	0.26 ± 0.10	
^{38}Cl	1.311	0.640	$PbCl_2$	0.15 ± 0.05	1.20 ± 0.40	1.20 ± 0.40
^{38}Cl	1.311	0.640	Au	o.09 ± 0.05	$1.00^{+1.5}_{-0.4}$	

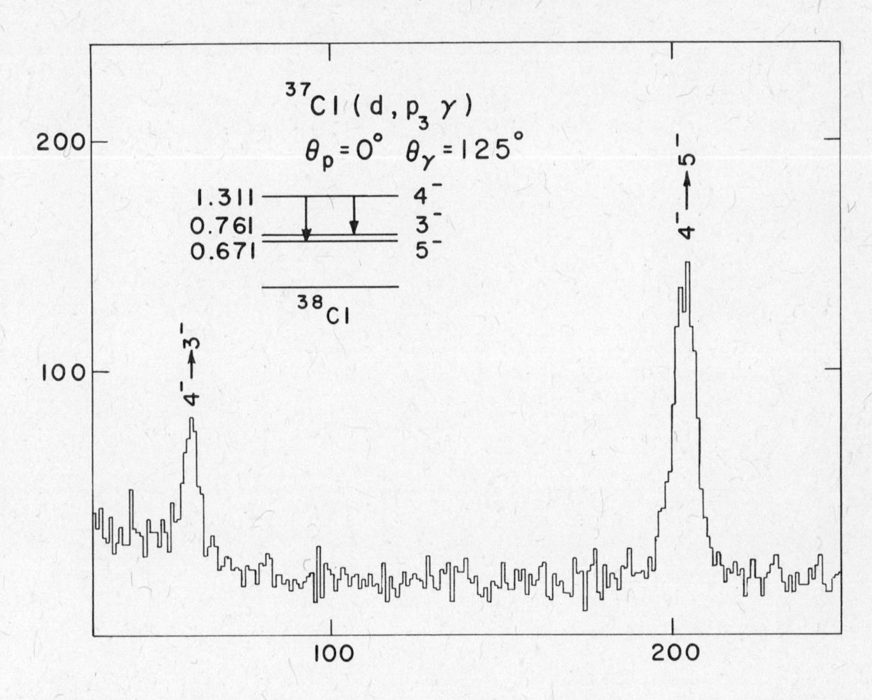

Fig. 13. Doppler-shift measurements in K^{40} via the $K^{39}(d,p)$ reaction at E_d = 3.5 MeV. The recoils were stopped in KI.

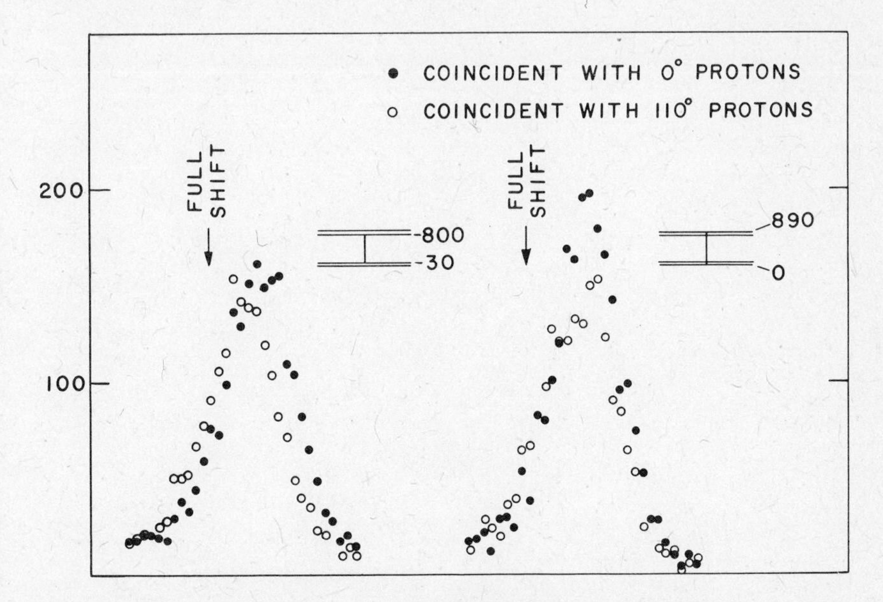

Fig. 14. Measurement of the branching ratio for the decay of the 1.311-MeV state in Cl^{38}.

Table V summarizes the results. Included is the 30-keV transition from the first excited state to ground in K^{40}. The lifetime of the 30-keV state in K^{40} has been measured electronically by three groups,[25-27] and the results all agree to within about 10%. The same $|M_{df}|^2$ should be found for all M1 transitions within a given multiplet, but in both K^{40} and Cl^{38} a discrepancy of a factor of at least 2.5 is found.

TABLE V

The quantity

$$|M_{df}|^2 \equiv [\langle\nu f_{7/2})||\vec{\mu}||(\nu f_{7/2})\rangle - (21/5)^{1/2}\langle(\pi d_{3/2})\rangle]^2$$

$$= \frac{2.99 \times 10^{-14}}{\tau E_\gamma{}^3(2J_f + 1)W^2(J_i J_f \tfrac{7}{2}\tfrac{7}{2};1\tfrac{3}{2})}$$

for various transitions. If the configuration were pure, $|M_{df}|^2$ would be a constant in each nucleus.

Transition	K^{40}	Cl^{38}
$3^- \leftrightarrow 2^-$	6.1 ± 1.5	16 ± 4
$4^- \leftrightarrow 3^-$	8.7 ± 0.8	2.6 ± 0.9[a]
$5^- \leftrightarrow 4^-$	3.2 ± 0.6	8 ± 3[a]
Calc. using free moments	6.24	6.24
Calc. using effective moments	8.09[b]	13.09[c]

[a]In Cl^{38} $\dfrac{|M_{df}|^2_{4\to5}}{|M_{df}|^2_{4\to3}} = 3.7 \pm 0.5.$

[b]Obtained using $\mu(Ca^{41})$ and $\mu(K^{39})$.

[c]Obtained using $\mu(Ca^{41})$ and $\mu(Cl^{37})$.

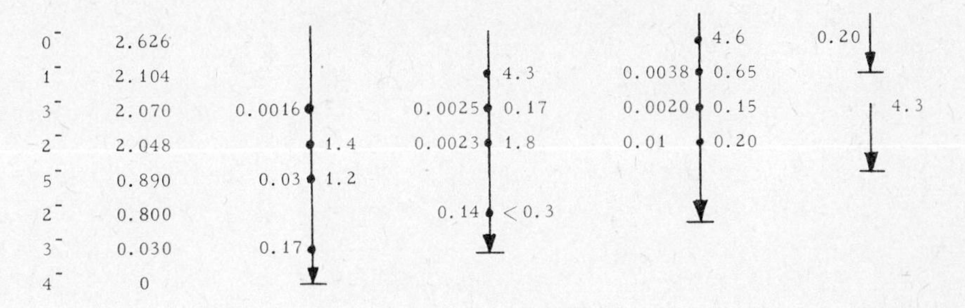

Fig. 15. Rates (Ref. 23) for transitions between negative-parity states in K^{40}.

Another set of measurements in K^{40} has recently been published[23] and are shown in Figure 15. The results for the second and third excited states are in very good agreement with those obtained at Argonne. It thus appears that there is very little chance that the discrepancy with theory can be attributed to experimental error.

Some remarks about the transitions in the $(\pi d_{3/2})$ $(\nu f_{7/2})$ quartet in Cl^{38} and K^{40} are in order. First of all, these are not particularly slow transitions. As a group, their speed is close to the high end of the M1 rate distribution (Figure 1). Secondly, no plausible E2 admixtures can rectify the situation. In fact, in K^{40} any E2 admixtures only make the discrepancy worse because the 30-keV transition is too low in energy to have a significant E2 component, while any E2 component in the 800-keV line has only the effect of further reducing the speed of the already too slow M1 component.

Another thing to be noted is that the expected leading impurities in the wave functions will have little effect on the M1 transition rates. The closest negative-parity states are predicted to be primarily of the $(\pi d_{3/2})$ $(\nu p_{3/2})$ and $(\pi s_{1/2})$ $(\nu f_{7/2})$ configurations. The fact that in K^{40} the M1 transitions from upper states to the ground-state quartet are quite slow supports this conjecture (Figure 15). Thus, on the one hand, the leading impurities will not have much effect and, on the other, effective moments are being used in order to take into account the effect of small admixtures of higher configurations. This should then be a very favorable case--probably as favorable as can be found. Yet the relative transition probabilities cannot be reproduced to within a factor of 2.5.

The poor agreement of the relative transition rates with theory was not expected and raises the question as to whether calculations using effective charges may be just as far off.

VI. REMAINDER OF THE sd SHELL

Between the two regions that have been discussed (i.e., between the collective region ending at about mass 27 and the region, which starts at about mass 38, in which states are constructed by starting with a spherical Ca^{40} core) is a region which includes about half of the shell. It seems to be a very difficult region to deal with. The presence of fast E2 transitions demonstrates that collective effects are important, but a rigid rotator does not fit the data well.

The difficulties are illustrated by a recent compilation[28] of static quadrupole moments that have been measured throughout the shell (Figure 16). From O^{16} to about mass[26], everything is very nice: the nuclei start out spherical and then show an almost constant prolate deformation between adjacent nuclei--frequently with changes in sign. The presence of significant deformations is confirmed by the fast E2 transitions, but instead of having a rigid rotator one seems to have a soft nucleus with a varying deformation. Bands are strongly mixed and it is very difficult to theoretically reproduce the observed data.

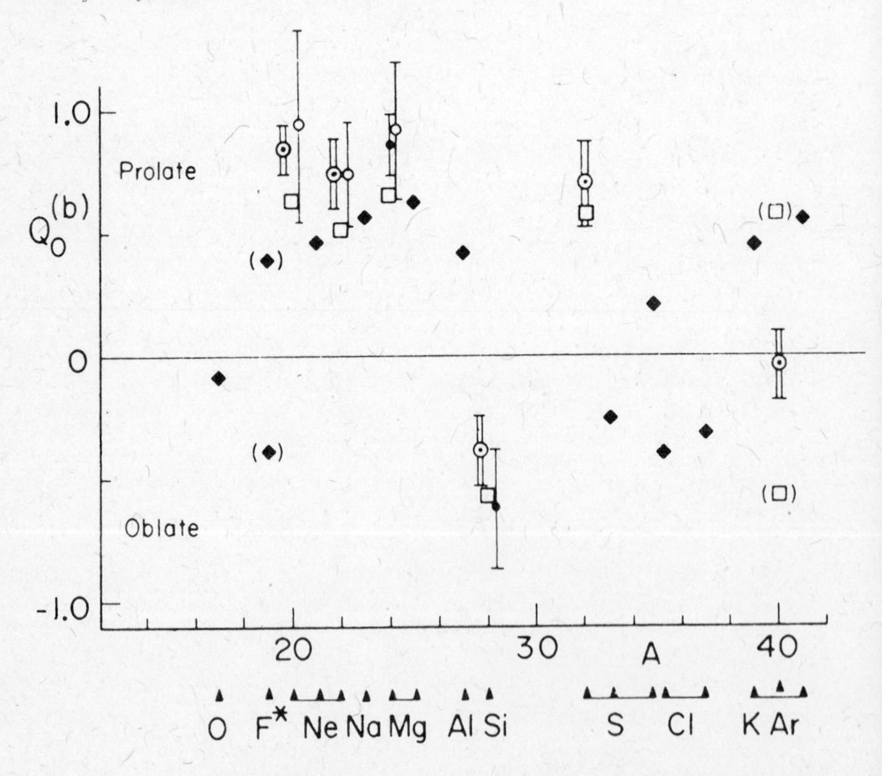

Fig. 16. Measured quadrupole moments given in Ref. 28.

A case in point is Si^{29}, a nucleus on which a large amount of data is available. The spins, parities, and lifetimes of the first seven excited states are all known.[29,30] The branching and mixing ratios in the decay of these states are also known, as are some properties of some of the higher states. From the spectroscopy alone (Figure 17), a rotational-band picture could probably be devised; but if one looks at the E2 transition rates, a simple picture just does not hold up. As an example, both low-lying $\frac{3}{2}^+$ states have strong E2 transitions to the ground state, but certainly both states cannot be members of the same band. Band mixing could be invoked, but it would be very difficult to do quantitatively.

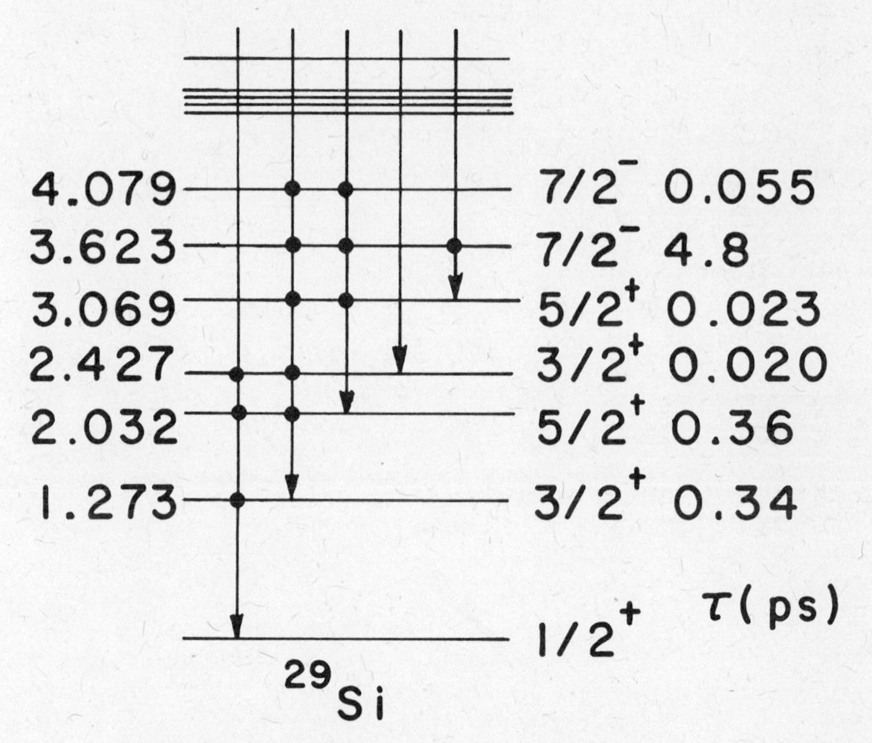

Fig. 17. Energy-level diagram of Si^{29}.

VII. SUMMARY

In summary, in the past couple of years there has been a great deal of new data on the lifetimes of transitions from bound states. A problem has arisen with the M2's. Several of these now are known, and the inhibition that has been established in the upper part of the periodic table does not seem to hold below A = 40. The fast M3 transitions in the A = 24 nuclei may or may not be a problem.

In terms of fitting the data to models, there is often very good agreement when a rigid-rotator model is used in the $19 \leq A \leq 26$ region. In several cases the relative transition rates can be calculated to within the accuracy of the experimental data which is, in places, as good as 30%. A glaring exception, however, is Na^{23}. At the ends of the shell, the spherical shell model should be valid; yet even when effective moments are used in K^{40} and Cl^{38}, the relative transition rates cannot be reproduced to within a factor of 2.5. This seriously limits what one can expect from calculations using effective charges and effective moments.

New data in this area are being generated constantly, and it can be expected that much of this information will help to fill in the picture outlined above. It can also be hoped that some of the new data will help lead to solutions of the problems that are still outstanding.

REFERENCES

1. S. A. Moszkowski, in *Alpha-, Beta-, and Gamma-Ray Spectroscopy*, edited by K. Siegbahn (North-Holland Publishing Co., Amsterdam, 1965), Vol. II, p. 865.

2. D. Strominger and J. O. Rasmussen, Nucl. Phys. 3, 197 (1957).

3. D. Kurath and R. D. Lawson, Phys. Rev. 161, 915 (1967).

4. P. M. Endt and C. van der Leun, Nucl. Phys. A105, 1 (1967).

5. L. A. Radicati, Phys. Rev. 87, 521 (1952).

6. G. Morpurgo, Phys. Rev. 110, 721 (1958).

7. A. E. Blaugrund, D. H. Youngblood, G. C. Morrison, and R. E. Segel, Phys. Rev. 158, 893 (1967).

8. D. H. Youngblood, R. C. Bearse, N. Williams, A. E. Blaugrund, and R. E. Segel, Phys. Rev. 164, 1370 (1967).

9. K. Sugimoto, Phys. Rev. 182, 1051 (1969).

10. E. K. Warburton, Phys. Rev. Letters 1, 68 (1958).

11. E. K. Warburton and J. Weneser, in *Isospin in Nuclear Physics*, edited by D. H. Wilkinson (North-Holland Publishing Co., Amsterdam, 1969), p. 173.

12. A. E. Litherland, in *The Structure of Low-Medium Mass Nuclei*, edited by J. P. Davidson (University of Kansas Press, Lawrence, Kansas, 1968), p. 92.

13. J. W. Olness, *ibid*, p. 192.

14. J. G. Pronko, C. Rolfs, and H. J. Maier, Phys. Rev. 186, 1174 (1969).

15. J. L. Durell, P. R. Alderson, D. C. Bailey, L. L. Green, M. W. Greene, A. N. James, and J. F. Sharpey-Schafer, Phys. Letters 29B, 100 (1969).

16. H. J. Maier, J. G. Pronko, and C. Rolfs, Nucl. Phys. A146, 99 (1970).

17. J. V. Kane, R. E. Pixley, R. B. Schwartz, and A. Schwarzschild, Phys. Rev. 120, 162 (1960).

18. R. C. Bearse, D. H. Youngblood, and R. E. Segel, Nucl. Phys. A111, 678 (1968).

19. W. A. Gerace and A. M. Green Nucl. Phys. A93, 110 (1967).

20. S. P. Pandya, Phys. Rev. 103, 956 (1956).

21. S. Goldstein and I. Talmi, Phys. Rev. 102, 589 (1956).

22. P. J. Twin, W. C. Olsen, and E. Wong, Phys. Letters 29B, 570 (1969).

23. R. Bass and R. Wechsung, Phys. Letters 32B, 602 (1970).

24. R. E. Segel, G. H. Wedberg, G. B. Beard, N. G. Puttaswamy, and N. Williams, Phys. Rev. Letters 25, 1352 (1970).

25. F. J. Lynch and R. E. Holland, Phys. Rev. 114, 825 (1959), and private communication.

26. D. W. Hafemeister and E. Brooks Shera, Phys. Rev. Letters 14, 593 (1965).

27. J. F. Boulter, W. Y. Prestwich, and B. Arad, Can. J. Phys. 14, 593 (1965).

28. K. Nakai, J. L. Quebert, F. S. Stephens, and R. M. Diamond, Phys. Rev. Letters 24, 903 (1970).

29. S. I. Baker and R. E. Segel, Phys. Rev. 170, 1046 (1968).

30. M. J. Wozniak, R. L. Hershberger, and D. J. Donahue, Phys. Rev. 181, 1580 (1969).

DISCUSSION

WARBURTON: I would like to say a few words about M2 transitions. First of all, let me repeat what you have said, Ralph, when you were talking about M3 rates, namely, that the single-particle units are rather arbitrary. Thus, the single-particle unit for, say, a $d_{3/2}$ to $f_{7/2}$ neutron transition is about 10 single-particle (Weisskopf) units, whereas E1 single-particle transitions have strengths which are typically only fractions of a single-particle unit. So when you make a comparison like that of your first slide you must bear this in mind and cannot conclude that there is no room for retardation of M2 rates as well as E1 rates. Secondly, I would like to enlarge on the similarity that Kurath and Lawson talk about between the E1 and M2 transitions -- they say that the space part is similar; therefore, that the selection rules that hold for E1 transitions will also hold in some respect for M2 transitions. It is also true that the isospin effects are somewhat similar. The moments that come into the M2 transitions, μ_n and μ_p, are opposite in sign and roughly the same in magnitude and when you take the combinations of them necessary to conserve isospin you often get cancellation. Likewise, in the E1 case you have effective charges of N/A and this retardation for the proton and neutron, respectively. And when you combine this so as to conserve isospin -Z/A it gives you about the same retardation effect as for M2 transitions, so that the similarity is quite strong. An example of M2 retardations in the s,d shell which just follows the Kurath-Lawson argument for isospin retardation is Ne^{21}. The $p_{1/2}$ hole state at 2.79 MeV decays to the 3/2+ and 5/2+ ground state, i.e., first excited state, and in calculating these rates you have to take the right combination of neutrons and proton matrix elements (it turns out to be twice the neutron and once the proton for a T = 1 core) and this gives you a factor of 10 retardation. Another factor of about 1/2 comes about because of the many-particle nature of the wave function so the M2 rate retarded by a factor of 20, but it doesn't look retarded because the single-particle unit is 10 Weisskopf units so we still end up with 0.5 Weisskopf unit.

SEGEL: I think in the Kurath and Lawson paper they do list retardation in terms of single particle units which are defined by the Moszkowski estimate and for greater than mass 40 they are all no greater than one percent of the Moszkowski single-particle unit. While here, they are of the order of 10 to 1.

ZAMICK: I want to talk about the M1's that you mentioned in K^{40} between say, a 3- and 4-state. I think that you can get the "anomalous" results by configuration mixing, for example, the basic configuration is $(f_{7/2}, d_{3/2}^{-1})$. By admixing configurations like $(f_{5/2}, d_{3/2}^{-1})$ or $(f_{7/2} d_{5/2}^{-1})$ you can alter the M1 transitions and maybe explain your data. This configuration mixing cannot be simulated by modifying the g factors.

SEGEL: Would it also effect the Pandya transformation?

ZAMICK: Probably not significantly.

SEGEL: I think one has to show how one can change one and not the other.

ZAMICK: If you just had an isolated $f_{7/2}$ nucleon you couldn't possibly have any $f_{5/2}$ admixture because it would violate angular momentum; 7/2, 5/2, I can't have in the same wave function, but if I bring the particle and hole together then I can get $f_{5/2}$ into the wave function, e.g., $(f_{5/2}, d_{3/2}^{-1})^{3-}$ and therefore this configuration mixing cannot be a term which is separately contained in the g factor of the $7_{7/2}$ particle or in the g factor of the $d_{3/2}$ hole.

SEGEL: That's all right. That is why I used effective moments.

ZAMICK: But this is a new effective moment which cannot be present separately in the $f_{7/2}$ or in the $d_{3/2}$ because neither of those have $f_{5/2}$ admixture. So possibly, (I am not going to guarantee it) such configurations might be small and might explain the M1's; of course, one has to do the calculations to verify that. I don't know if it will.

SEGEL: It will certainly have to be explained by mixing in other states, but I think it is a little disburbing that one has to.

ZAMICK: Say one only needs 2% of admixture as above. It's not impossible *a priori*. I don't know though.

KUO: I would like to ask you the following. I have sent you some wave functions which included configuration mixtures for the $d_{3/2}^{-1}$, and so forth. I would just like to know now what would the results be if one brings in the configuration mixtures?

SEGEL: I didn't do that. Your leading impurities are just about what one would expect, the neutron in the p state and the $s_{1/2}$ proton hole. However, I thought the point of the effective moments was that the small bits of other configurations could be taken into account just as with effective charges. I have no doubt if one writes the wave function out completely one can find the conbination that will reproduce the transition rates, but it does throw some doubt on how well you can do when you don't do that, but use effective moments and effective charges instead.

ZAMICK: What I mentioned before, e.g., $(f_{5/2}d_{3/2}^{-1})$ mixing, is a counter example of the statement that you can take separate g factors for the $f_{5/2}$ and the $d_{3/2}$ and then do the calculations.

SEGEL: No, no, but the effective moments that were used did not have to be, say the moments of K^{39} and Ca^{41}. You could use whatever you want and still the relative transition rates are fixed. The disturbing thing was not that you couldn't produce the absolute rates, but that you couldn't reproduce the relative rates. You have to have different effective moments for the four states.

ZAMICK: Even so, this particular process cannot be represented by using arbitrary effective moments for $f_{7/2}$ and $d_{3/2}$. It is a process in which the $f_{7/2}$ and $d_{3/2}^{-1}$ get tangled up, they both contribute simultaneously to the configuration mixing.

SEGEL: But wouldn't that always be true? For anything you do and try to reproduce an effective moment or an effective charge you would always have that problem because you get it by mixing in different configurations.

MacDONALD: In the case of Ca^{40}, one has a similar quartet of negative parity levels, which, to first approximation, are a $(d_{3/2}^{-1},f_{7/2})$ configuration. It is not until one does a calculation which includes RPA and deformed components that experiment and theory can be brought into agreement for 9 E2 transitions between negative parity levels.

SEGEL: Yes, but that was just the point here, when we don't have a lot of parameters how well we fit?

MacDONALD: Right, what you are saying in fact is that one has to do all of this before you even get close in some cases.

SEGEL: Well, I am not sure that is what I am saying.

HOROSHKO: I want to mention that one has to be very cautious in drawing conclusions from the agreement that one obtains from the Pandya transformation, conclusion with respect to the purity of the states. This transformation seems to work in the cases where one might not expect it to work; for example, we find that the $(d_{3/2})^2$ multiplet in Cl^{34} can be related quite well to states in Cl^{36} with this transformation even though some of these states are expected to be only about 50% pure.

SEGEL: Yes, I would certainly agree that one could re-phrase the conclusion by saying that the model isn't very good and that even though it works for the Pandya transformation the fact that it doesn't reproduce the transition rate shows that it isn't a very good model. On the other hand, if it isn't, then where will the simple model be good and where can you trust calculations of transition rates based on a simple model and that is why I tend to draw the conclusion that this shows that there is a real limitation to how well one can do in calculating transition rates. It might not apply how well you can calculate energy level spectra.

HOROSHKO: One often finds that the Pandya transformation works quite well for energies but not for transitions.

HALBERT: Perhaps this is a way of saying it. The energy operator is simple for a small shell-model space, whereas the effective magnetic moment operator for that small space would be more complicated than you could get by simply using effective single-particle moments in the ordinary magnetic moment operator.

HANNA: I wanted to ask you exactly what your criterion is for saying that the transition rates for the rotational levels come out well. I think you get something like 30% agreement and you can't do better because of the experimental error. But in a way it seems that hidden in this agreement may be exactly the same kind of trouble that you have when you are looking at the shell model transition rates at the ends of the shell, because of course, in the rotational model you are measuring just a very simple thing -- you are measuring the E2 rate for a very dominant motion and any deviation due to the shell model would be masked out.

SEGEL: Where we think the rigid rotator is applicable, the relative transition rates agree to about 30%, which is the quoted experimental error, while here the disagreement is about a factor of 2-1/2, so there certainly is a quantitative difference. I agree that the rotational picture seems to be the one, that where it's applicable it works very well, and one can consider that an empirical fact, and perhaps not a surprising one.

HANNA: Let's put it another way. I think that if you were to find a 20% disagreement in the rotational rates you ought to be worried about it. I think it might well be as serious from the viewpoint of nuclear structure as a two-to-one disagreement in say, Ca^{40}.

HALBERT: There is some experimental evidence, having to do with the quadrupole moments of 2+ states in comparison with B(E2) values for 2+ → 0+ transitions, indicating there is some breakdown of the rotational model. The rotational model gives you a value for the ratio of $[B(E2: 2+\rightarrow 0)]^{1/2}$ to Q(2+), for a K = 0 band. There is some controversy about the experimental information, but it does seem that in three or four s,d-shell nuclei for which measurements have been made, this ratio is smaller than the rotational value by about 30%. Probably, no one measurement can be trusted to more than 30%, but it's interesting that there seems to be a trend, that the measurements for four different nuclei show this 30% violation in the same direction.

SEGEL: I think both types of data were on that slide that I showed and I do have the reference in Phys. Rev. Letters fairly recently, at least the impression that I got of the person who wrote the letters that he thought the experimental errors were large enough that it is really in agreement.

HALBERT: The four nuclei included Ne^{20}, Ne^{22}, and Mg^{24}, as I remember it.

SEGEL: No, they are the re-orientation measurements, and these are very difficult.

WILDENTHAL: First of all, it is much like several other fields. The measurements are pretty good, that is, the quoted errors are down to around ten percent but the theory with which people extract the model-testing numbers still causes trouble; so the error bars are now quite small, but one still wonders about the numbers that are extracted. It takes us a little bit afield, but with the shell model we can reproduce ground state quadrupole moments, within their error bars (which again involve a different kind of theoretical uncertainty), but we are rather consistently 30 to 40% low versus the re-orientation measurements.

CASTEL: Well, I think your comment on this comparison between theoretical and experimental microscopic nature of the rotational band is really interesting because in Si^{28} or Mg^{24}, for instance, you can see that the quadrupole moments of all the excited states calculated in the rotational model would decrease as you go up the rotational band. In Mg^{24} for instance, it's really very interesting and in Si^{28} I see the ratio of the quadrupole moments of the 4+ 2+ states in this rotational model is already .5 and I don't think there is any way of explaining this by band mixing because the oblate band is really much too high. There is a very little overlap between the two solutions. We have tried, for instance, to project the Hartree-Fock to try to minimize the after projection for every one of these states of the rotational band, and it is interesting to see that the quadrupole moments really decrease if you make sure that you get bigger binding energy by projecting, by varying after projection. So I think maybe this answers part of your comparison.

III.A. HARTREE-FOCK CALCULATIONS IN THE 2s-1d SHELL*

S. J. Krieger

University of Illinois at Chicago Circle, Chicago, Illinois,
and
Oak Ridge National Laboratory, Oak Ridge, Tennessee

I. INTRODUCTION

The Hartree-Fock method has gained immense popularity as a means of calculating a variety of nuclear properties for a broad range of nuclei. The method has been used to calculate binding energies, rms radii, and nuclear shapes for a number of nuclei, and, when supplemented by angular momentum projection, has also provided an extremely useful approximation to exact diagonalization within a shell-model space. Calculations of the latter type have helped to provide a connection between macroscopic and microscopic descriptions of nuclear deformation. In addition, however, to the calculational benefits derived from Hartree-Fock, there has also been emerging a body of knowledge concerning the nucleon-nucleon interaction itself. Before turning to the main topic of this talk, I would like to briefly discuss the importance of this body of knowledge, and divert attention for a moment to the present theoretical understanding of the nucleon-nucleon interaction.

From field theory we have learned a great deal about the interaction of low momentum transfer. While from the two-body scattering data the *on-shell* two-body T-matrix has been determined with reasonable unambiguity. Note however, that extremely little is known of the *off-shell* T-matrix. (In terms of potential theory the corresponding statement is that existing data are unable to decide between an infinite class of potentials which have remarkably diverse short and even intermediate range behavior.) The point that I wish to make is that further progress in our understanding of the interaction is most likely to come from a critical analysis of microscopic calculations of nuclear effects. That is by including what is known about nucleon interactions in nuclear calculations, one can not only provide less ambiguous nuclear predictions, but hopefully can provide criteria for choosing between otherwise equally valid interactions.

---

*Research sponsored by the University of Illinois at Chicago Circle, the U. S. Atomic Energy Commission under contract with Union Carbide Corporation, and the Oak Ridge Associated Universities.

As a not particularly recent example of how such information may become available, consider the empirical evidence derivable from nuclear structure studies which supports the statement that the "real" nucleon-nucleon interaction, whatever it may be, must be too strong to be treated in the Hartree-Fock approximation. Specifically, it is not possible within the Hartree-Fock approximation to obtain the correct single particle levels *and* the correct size and total binding energy of a nucleus.

The empirical nature of the evidence can be understood if one recalls the result that the binding energy, in the Hartree-Fock approximation, is simply one-half of the difference of the separation energies and the kinetic energies of the occupied orbitals. Thus, if one takes the separation energies from experiment, and computes the kinetic energies using the known rms radius of the nucleus, one obtains, for example in the case of O^{16}, but two-thirds of the experimental binding. The conclusion is that one cannot make the interaction arbitrarily soft, and as a corollary, one must deal with the higher order corrections.

Even in the face of the above discussion, let me state that I believe that it is important to continue the search for a "soft" potential suitable for Hartree-Fock. In the first place the calculations performed with such a potential can greatly contribute to our understanding of nuclear spectra. Secondly, until the uncertainties of the nucleon-nucleon interaction are better resolved, such a search can only contribute to an understanding of nucleon correlations--albeit through their omission.

The calculations which I wish to discuss today were carried out with the explicit intention of testing how well one could do, utilizing a rather smooth interaction, in providing descriptions of binding energies, density distributions, and nuclear shapes, for the light nuclei. In an effort to understand the validity of some of the approximations used in previous computations, I have used a general two-body potential which, although developed especially for use in Hartree-Fock calculations, may yet be described in some sense as having been derived from the free two-nucleon data. Thus the calculation is more fundamental than one which uses a *purely* phenomenological potential, and as such may be viewed as a standard with which to check, for example, the validity of using an effective one-body spin-orbit interaction. Moreover, the present calculation, in addition to varying the orbits of all particles, utilizes a basis for which the effects of truncation are minimal so that the deformations found are not limited by restrictions on the space of the Hartree-Fock wave function. As I feel that this is an important point, I would like to discuss it more fully.

II. SYMMETRIES OF THE HARTREE-FOCK WAVE FUNCTION

Because it is necessary in practice to truncate the basis in which the Hartree-Fock wave function is expanded, the choice of basis is *not* arbitrary. In particular, as we wish to study nuclei which may not be axially symmetric, it will clearly be advantageous to expand the orbitals in a Cartesian oscillator basis

$$|\lambda\rangle = \sum_{n_x n_y n_z \sigma} C^{\lambda}_{n_x n_y n_z \sigma} |n_x\rangle |n_y\rangle |n_z\rangle |\chi^{\frac{1}{2}}_{\sigma} \tag{1}$$

where

$$\langle x|n_x\rangle = [\pi^{\frac{1}{2}} 2^{n_x} n_x! b_x]^{-\frac{1}{2}} H_{n_x}(x/b_x) e^{-x^2/2b_x^2} \tag{2}$$

and H_{n_x} is the Hermite polynomial given below in terms of the generating function

$$e^{-s^2 + 2s\xi} = \sum_{n=o}^{\infty} \frac{s^n}{n!} H_n(\xi). \tag{3}$$

The advantage of this choice of basis is that the nuclear deformation can be incorporated directly into the basis. That is, by proper choice of the harmonic oscillator parameters b_x, b_y, b_z we can accelerate the convergence of the calculation with dimensionality. Stated differently, we can produce deformations with a set of basis functions for which $n_x + n_y + n_z \leqslant 2$ (i.e. the 2s-1d shell in the case $b_x = b_y = b_z$) that a calculation utilizing a basis of spherical oscillator functions could not produce, even if it were to include perhaps the 2p-1f and 1g shells. The Cartesian basis offers a similar advantage over a cylindrical basis in the study of tri-axial nuclei.

In order to minimize the number of two-body matrix elements which must be computed, it is desirable to introduce the expected symmetries of the Hartree-Fock wave function directly into the basis. Although this procedure is extremely attractive from the point of view of making the computation tractable, it is not without its drawbacks. For if we construct the basis in such a manner as to make the occupied orbitals $\{\lambda\}$ invariant under any symmetry operation which commutes with the many-body Hamiltonian H, then the one-body Hartree-Fock Hamiltonian $h(\lambda)$ will also be invariant under the symmetry operation. Thus, the symmetry will be propagated from iteration to iteration of the Hartree-Fock equations, with the result that an energetically more favorable solution of lower symmetry may, perhaps, be missed.

We have allowed general ellipsoidal deformation subject to the following additional symmetries: (1) The orbitals are assumed to have good parity, (2) the Hartree-Fock wave function is assumed to be invariant under time reversal, (3) it is invariant under rotations in isospin space. Note that assumptions (2) and (3) together require that each spatial state be fourfold occupied. The validity of assumptions (1) and (2) has been tested for a number of spherical and deformed nuclei by the M.I.T. group and in all cases, for reasonable interaction strengths, the symmetries have been found valid. For the light nuclei under consideration, neglect of the Coulomb force is a reasonable approximation.

Because of the above symmetries, the basis states (compare Eq. (1)) take the form

$$|\lambda\Pi R_z^+\rangle = \sum_{n_x n_y n_z} (i)^{\Pi_y} a^{\lambda}_{n_x n_y n_z} \Big[\frac{1+(-1)^{\Pi+\Pi_z}}{2} |n_x\rangle|n_y\rangle|n_z\rangle\chi^{\frac{1}{2}}_{\frac{1}{2}} + (-1)^{\Pi_x} \frac{1-(-1)^{\Pi+\Pi_z}}{2} |n_x\rangle|n_y\rangle|n_z\rangle\chi^{\frac{1}{2}}_{\frac{1}{2}}\Big] \tag{4}$$

$$|\lambda\Pi R_z^-\rangle = T|\lambda\Pi R_z^+\rangle = \sum_{n_x n_y n_z} (i)^{\Pi_y} a^{\lambda}_{n_x n_y n_z} \Big[\frac{1-(-)^{\Pi+\Pi_z}}{2} |n_x\rangle|n_y\rangle|n_z\rangle\chi^{\frac{1}{2}}_{\frac{1}{2}} + (-)^{\Pi_x} \frac{1+(-)^{\Pi+\Pi_z}}{2} |n_x\rangle|n_y\rangle|n_z\rangle\chi^{\frac{1}{2}}_{-\frac{1}{2}}\Big] . \tag{5}$$

where $\Pi_i = 0$ if n_i is even and $\Pi_i = 1$ if n_i is odd ($i = x,y,z$). ($|\lambda\Pi R_z^+\rangle$ transforms under $R_z(\pi)$ as a particle of spin + 1/2, while $|\lambda\Pi R_z^-\rangle$ transforms as a particle of spin - 1/2.) R and T are the operators which generate rotations and time reversal, respectively. The coefficients $a^{\lambda}_{n_x n_y n_z}$ are real as will be all matrix elements of the Hartree-Fock Hamiltonian between the basis states. The fact that $T|\lambda\Pi R_z^{\pm}\rangle = \pm\,|\lambda\Pi R_z^{\mp}\rangle$ facilitates the evaluation of expectation values of operators, e.g. $\langle HF|Q_{20}|HF\rangle$ or $\langle HF|J|HF\rangle$.

Due to the above symmetries of the Hartree-Fock wave function, the Hartree-Fock Hamiltonian must necessarily take the form depicted in Figure 1. Because the orbitals have been assumed to have good parity, the matrix divides into two blocks which operate on the even parity states and the odd parity states, respectively. Because of the assumed invariance under time reversal, each of the two blocks further divides into two identical blocks. That the blocks are identical is apparent, for the eigenstates of the blocks, the states R_z^+, R_z^-, are connected by an operator T which commutes with the Hamiltonian, and are thus degenerate. Up to this point in our discussion we have suppressed mention of the isospin quantum number. To be specific, the above array may be thought of as operating upon the neutron states. Then, because of the neglect of the Coulomb force, there is an identical array which operates upon the proton states, and each spatial state *is* fourfold occupied, as stated earlier.

The benefit of the above reduction lies not in the fact that we need only diagonalize two small arrays, e.g. Π even R_z^+ and Π odd R_z^-, but rather that in order to calculate these arrays we need only the two-body matrix elements $\langle n_\alpha \Pi_\alpha R_z^\alpha, n_\beta \Pi_\beta R_z^\beta\ SM|V_T|n_\gamma \Pi_\alpha R_z^\alpha, n_\delta \Pi_\beta R_z^\beta\ SM'\rangle$ listed in Table I. Further, the matrix elements of class 7 in Table I are clearly identical to those of class 3, while the matrix elements of class 8 are, through time reversal symmetry, equal to those of class 4. Thus, only the matrix elements of classes 1 through 6 need be computed. This simplification is what makes the computation tractable. For even with the introduction of the symmetries, we must calculate 72,010 (independent) matrix elements for the case $n_x + n_y + n_z \leq 3$.

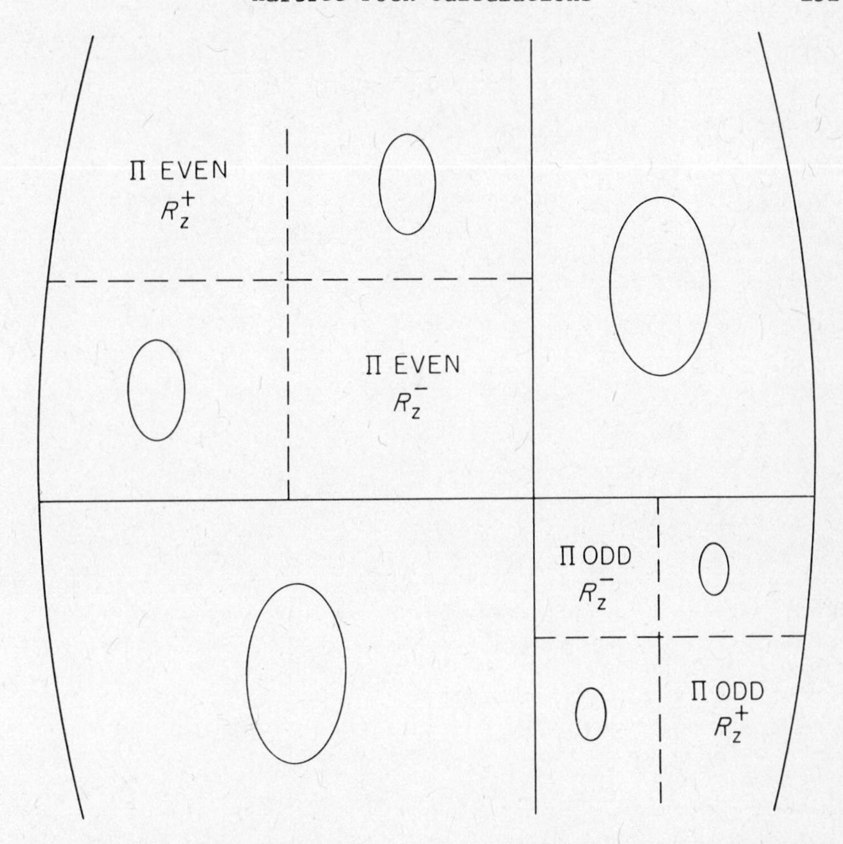

Fig. 1. The HF Hamiltonian in the oscillator representation.

III. THE TWO-BODY INTERACTION

In this work we use a recently developed velocity-dependent potential which was especially derived for use in Hartree-Fock calculations. The explicit form of the interaction is

$$(m/\hbar^2)U_j = V_j^c(r) + \frac{p^2}{\hbar^2}\,\omega_j(r) + \omega_j(r)\frac{p^2}{\hbar^2} + V_j^T(r)S_{12} + V_j^{LS}(r)\vec{L}\cdot\vec{S} \quad (6)$$

where m is the nuclear mass and r and p represent the relative coordinate and momentum of the interacting nucleons. $\vec{L}$ and $\vec{S}$ are the orbital angular momentum and spin angular momentum operators and S_{12} is the tensor

TABLE I

Two-body matrix elements Classified by Symmetry Group

Class	Π_α	R_z^α	Π_β	R_z^β
1	Even	R_z^+	Even	R_z^+
2	Even	R_z^+	Even	R_z^-
3	Even	R^+	Odd	R_z^-
4	Even	R_z^+	Odd	R_z^+
5	Odd	R_z^-	Odd	R_z^-
6	Odd	R_z^-	Odd	R_z^+
7	Odd	R_z^-	Even	R_z^+
8	Odd	R_z^-	Even	R_z^-

operator. The index j labels the four spin-parity states of the two-nucleon system, SE, SO, TE, TO. (No spin-orbit interaction has been included in the TE state.) The radial dependence of all terms is Gaussian; however, the tensor term is additionally multiplied by r^2: $V_j^T(r) = -A_j(r^2/2\alpha_j^2)\, e^{-(r^2/2\alpha_j^2)}$. With this choice of radial dependence, the interaction is separable in Cartesian coordinates and the computation of two-body matrix elements is thus greatly facilitated.

The potential saturates nuclear matter in the Hartree-Fock approximation at $k_F = 1.38\ \mathrm{fm}^{-1}$ with an energy $E_0 = -15$ MeV. The second-order Goldstone correction to the energy, calculated in the effective mass approximation, is -3.2 MeV. The low-energy scattering produced by the potential can be summarized as follows. The accepted values for the scattering lengths and effective ranges are fitted to within 5%, and the deuteron binding energy and quadrupole moment are fitted to within 10%. However, the percentage d state in the deuteron is calculated to be 2% which is much lower than the accepted value of approximately 6%. Complete details concerning this potential are given in Nestor *et al.*[2] For calculations of the closed-shell nuclei with this potential, see Tarbutton and Davies[3].

IV. SOLUTION OF THE HARTREE-FOCK EQUATIONS

Because we are dealing with non-spherical nuclei, we expect to find a multiplicity of solutions to the Hartree-Fock equations, corresponding to various deformations. There are two main methods which one may use to explore the various solutions. The first method utilizes the fact that the Hartree-Fock equations are to be solved by an iterative technique. We pick an initial density matrix, evaluate the Hartree-Fock potential Γ, diagonalize the Hartree-Fock Hamiltonian $h = T + \Gamma$, recompute Γ with a new density matrix determined by filling the orbitals of lowest energy obtained in the diagonalization of h and iterate until the sequence converges. Thus, all possible solutions should be found by simply exhausting all of the ways in which one can choose the initial density matrix. An alternative method is to "force" the nucleus to change deformation by placing it in an external (quadrupole) field. If the external field is large compared to the self-consistent nuclear field, then the nuclear shape will be determined by the external field and the nucleus can be driven (discontinuously, due to shell structure) from one solution to another. This latter method is particularly useful in gaining insight into the nature of excited states of collective nature, since it enables one to map the energy surface in the vicinity of a true solution. For the present, we restrict ourselves to a discussion of the solution at the true minima and adopt the first method outlined above. The Cartesian basis, as it turns out, is once again a rather natural one in that it is fairly easy to pick intuitively initial density matrices, which then converge to the desired results.

We then proceed by choosing a density matrix,

$$\rho_{n_2\sigma_2;n_1\sigma_1} = \sum_{\lambda \text{ occupied}} a^{\lambda}_{n_2} a^{\lambda}_{n_1} \tag{7}$$

compute the HF potential

$$\Gamma_{n\sigma;n'\sigma'} = \sum_{n_1\sigma_1 n_2\sigma_2 TS} \frac{2T+1}{2} C^{\frac{1}{2}\frac{1}{2}\,S}_{\sigma\sigma_1 M} C^{\frac{1}{2}\frac{1}{2}\,S}_{\sigma'\sigma_2 M'} \langle nn_1M|U_{ST}|n'n_2M'\rangle_A \times \rho_{n_2\sigma_2;n_1\sigma_1} \tag{8}$$

add to it the kinetic energy matrix

$$t_{n\sigma;n'\sigma'} = \langle n\sigma|\frac{\vec{p}^2}{2m}|n'\sigma'\rangle$$

to form the HF Hamiltonian and iterate as described above.

V. RESULTS

The results presented here use a basis of oscillator functions with $n_x + n_y + n_z \leq 2$. No significant improvement is obtained by including the set of functions with $n_x + n_y + n_z = 3$. Thus, for example, if we minimize the ground-state energy of Si^{28} using the lower dimensionality and then, without further change of the oscillator parameters, increase the basis, we find less than a 1% change in the total binding energy, single-particle energies, and nuclear size and shape.

Our results are in general agreement with the results of Ripka[4] and Stamp[5]. Thus, we find that the nuclei Ne^{20}, Si^{28}, and Ar^{36} are axially symmetric with Ne^{20} prolate, and Si^{28} and Ar^{36} oblate, while Mg^{24} and S^{32} are triaxial. Although the shapes agree with the general arguments put forth by Banerjee *et al.*,[6] two of the nuclei, Si^{28} and S^{32} possess solutions of different symmetry which in each case lie less than 4 MeV above the lower energy solution.

Table II lists some of the results of the study. In addition to total binding energies and *rms* radii, we give the (intrinsic) electric quadrupole moments Q_{20}, Q_{22} as defined by

$$Q_{2m} = \sum_{\text{protons}} \sqrt{(16\pi/5)}\langle r^2 Y_{2m}\rangle, \tag{9}$$

and also the expectation value of the angular momentum $\langle J^2\rangle$. As may be seen, the nuclear sizes are in reasonable agreement with experiment. Using the adiabatic approximation, we have compared the intrinsic quadrupole moment of Ne^{20} with that obtained in a shell model calculation and find our result of 44 fm^2 in reasonable agreement with the extracted value of 49 fm^2. The shell model calculation used an effective charge of ½e.

TABLE II

Nuclear Properties. This table lists the binding energies, root mean square radii, electric quadrupole moments, and angular momenta for the ground state of the nuclei studied. The experimental radii are taken from Reference 7.

	B.E. (MeV)	r_{rms} (fm)		Q_{20} (fm^2)	Q_{22} (fm^2)	$\langle J^2 \rangle$
		Thy.	*Exp.*			
He^4	3.30	1.84	1.67	0	0	0
Be^8	12.0	2.32		39.0	0	10.5
C^{12}	37.4	2.45	2.42	-17.3	0	8.77
O^{16}	76.3	2.53	2.71	0	0	0
Ne^{20}	95.3	2.77		44.1	0	20.0
Mg^{24}	117.	2.82	2.98	39.3	-7.9	23.6
Si^{28}	157.	3.00	3.04	-63.1	0	29.7
S^{32}	190.	3.12	3.12	-56.1	17.4	27.0
Ar^{36}	233.	3.16		-49.8	0	16.3
Ca^{40}	283.	3.20	3.50	0	0	0

The binding energy per particle is plotted vs. $A^{-1/3}$ in Figure 2. Note that the deformed nuclei lie above the straight line which is drawn as a fit to the spherical nuclei O^{16} and Ca^{40} and nuclear matter. Thus, higher-order correlations omitted in Hartree-Fock appear to be of more importance for the open-shell nuclei[8].

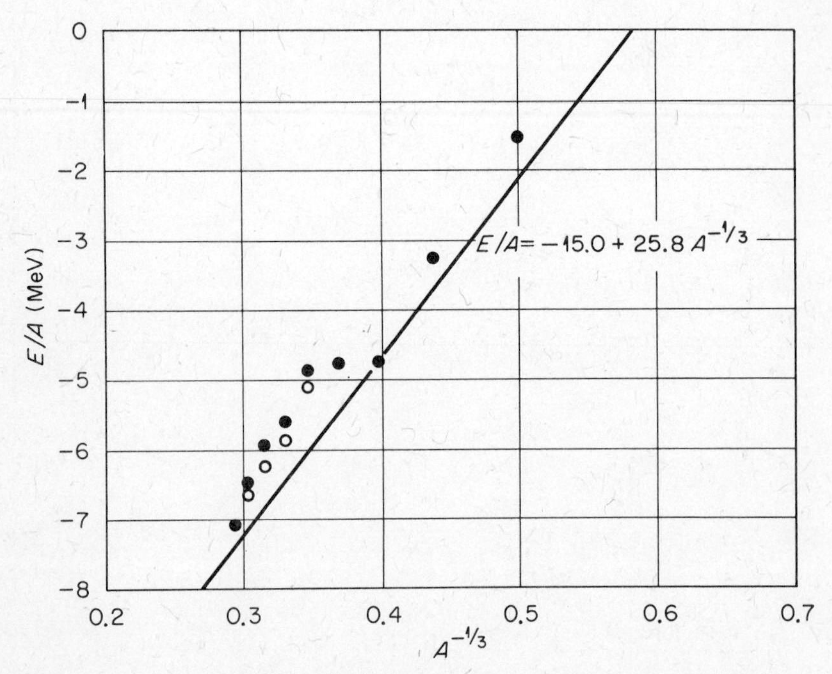

Fig. 2. Binding energy per nucleon. The open circles indicate the approximate energies which would be obtained upon projection, i.e.

$$E_{J=0} \simeq E_{HF} - \frac{\langle J^2 \rangle}{2I_{exp}} .$$

The straight line has been drawn as a fit to the spherical nuclei and nuclear matter.

In Figure 3 we compare the single-particle energies of the ground-state Hartree-Fock solutions with those of Ripka[4] and Stamp[5]. Where comparison is possible, there is qualitative and, in most cases, semi-quantitative agreement. The levels of the axially symmetric nuclei are compared with experiment in Table III, where the agreement is again reasonable.

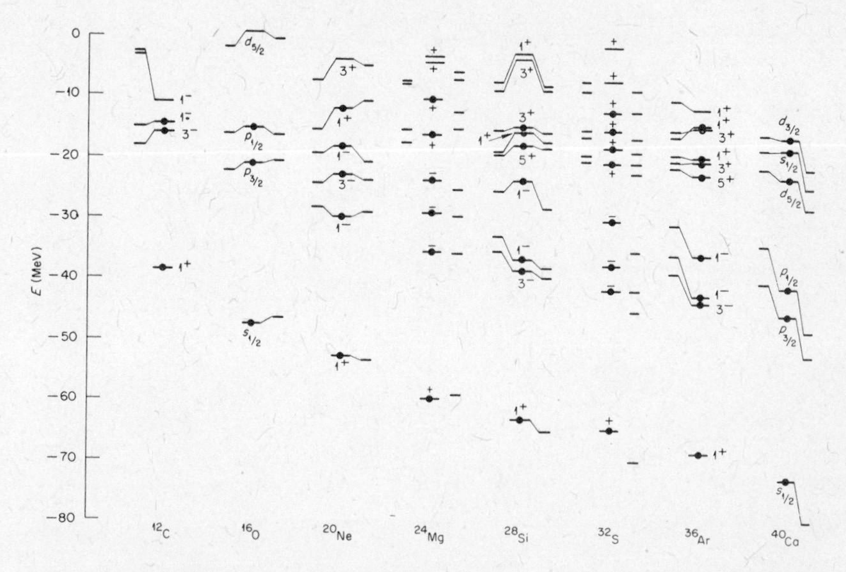

Fig. 3. Single-particle energy levels of the ground-state HF solutions. The central column for each nucleus contains the results of the present calculation. The levels of the axially symmetric nuclei are labeled by the z-component of angular momentum k and parity π as $(2k)^{\pi}$; the levels of the tri-axial nuclei are labeled by the parity alone. Occupied orbitals are indicated by a dot. The results are to be compared with the calculations of Ripka[4] (left column) and Stamp[5] (right column).

Further comparison with the results of Ripka[4] is made in Table IV where it may be seen that, with the exception of the excited (prolate) state solution of Si^{28}, there is rather remarkable agreement on nuclear shapes. The contribution of the valence particles to the electric quadrupole moment and to $\langle J^2 \rangle$ is plotted in Figure 4, where it is seen that the core contribution to Q_{20} is somewhat less than the $\sim$30% found in calculations using a basis of spherical oscillator functions. In order to obtain a more graphic representation of the varied equilibrium shapes, we have plotted in Figures 5a and 5b, equidensity surfaces in the plane containing the symmetry axis for the oblate and prolate solutions of Si^{28}. The corresponding wave functions are given in Table V. The results for the prolate solution, B.E. = 153 MeV, r_{rms} = 3.06 fm, Q_{20} = 76.1 fm^2, and $\langle J^2 \rangle$ = 28.1, may be compared with the corresponding numbers for the oblate solution which are given in Table II. The near identity of the rms radii for the two solutions does not in itself provide a very

TABLE III

Single-particle binding energies in MeV. With the exception of the $d_{5/2}$ level which, in the deformed nuclei is almost pure $d^{5/2}_{5/2}$, the energies of the deformed nuclei are labeled by $(2k^{\Pi})$; e.g., 3^- indicates a state of negative parity with z component of angular momentum 3/2. The experimental results are taken from Reference 9. For Ca^{40} the underlined energies represent neutron levels; all other experimental energies are for protons. For the deformed nuclei we have suggested a comparison with the state(s) which contain a significant amount of the single-particle strength; no attempt has been made to compute the center of gravity of the levels.

	O^{16}		Ne^{20}		Si^{28}		Ar^{36}		Ca^{40}	
	Thy.	*Exp.*	*Thy.*	*Exp.*	*Thy.*	*Exp.*	*Thy.*	*Exp.*	*Thy.*	*Exp.*
$1s_{1/2}$	47	44							73	≃ 75
$1p_{3/2}$	21	19	(3^-)23	19,18	(3^-)38	36			46	
$1p_{1/2}$	15	12	(1^-)30,18	12	(1^-)36,23	28			41	≃ 32
$1d_{5/2}$					17	17	22	20	23	22 16
$2s_{1/2}$					(1^+)16,2	13			18	18 8
$1d_{3/2}$									16	16 8

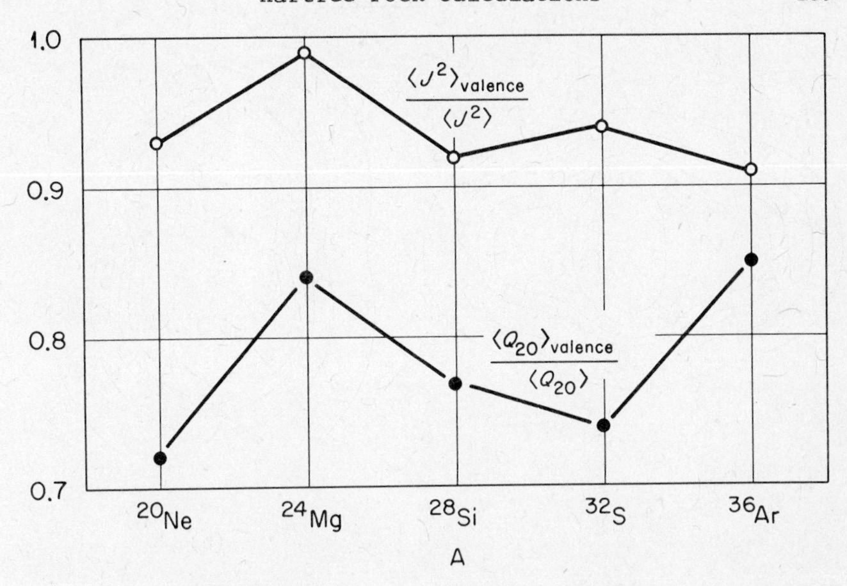

Fig. 4. Contribution of 'valence' particles to the expectation value of the angular momentum and quadrupole moment. The nuclei Ne^{20}, Mg^{24}, Si^{28} are considered as having 4, 8, and 12 valence particles, while S^{32} and Ar^{36} are considered as having 4 holes and 8 holes, respectively.

TABLE IV.

Comparison of nuclear deformations. The second column contains the results of the present work, while the third column contains the (major-shell mixing) results of Ripka[4].

Nucleus	$\frac{\langle 2z^2-x^2-y^2\rangle}{\langle x^2+y^2+z^2\rangle}$	
C^{12}	-.48	-.47
Ne^{20}	.57	.59
Si^{28} oblate	-.50	-.49
Si^{28} prolate	.58	.75
S^{32} prolate	.33	.35
A^{36}	-.28	-.24

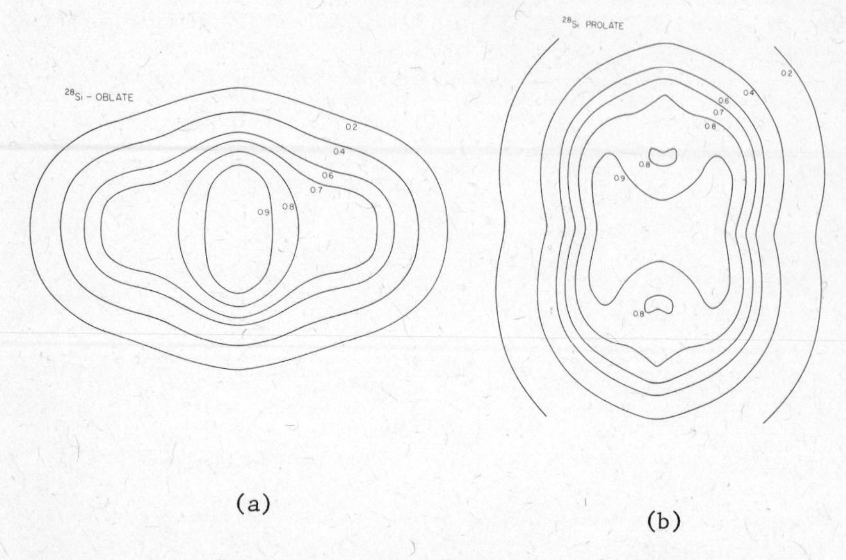

Fig. 5. Cross sections of equidensity surfaces cut by a plane containing the symmetry axis for the oblate (a) and prolate (b) solutions of Si^{28}. The numerical values refer to the fraction of maximum density: 0.26 fm^{-3} and 0.20 fm^{-3}, respectively for the two solutions.

valid test of the volume conservation hypothesis for the value of this parameter is very heavily influenced by the nuclear *surface*. In an effort to obtain a more meaningful comparison, we have calculated the volumes of the oblate, prolate, and spherical states of Si^{28} as defined by

$$V = \int \underline{d}^3 r \theta[\rho(\underline{r}) - .2\rho_{max}],$$

i.e., the volume in which the density is no less than 20% of the maximum density. We find that the volumes defined in this manner differ by less than 10%. A more accurate determination of this volume might reduce the variance to 5%.

VI. CONCLUSIONS

In looking over various efforts, nearly all nuclear calculations yield rather similar results for the energy levels and equilibrium shapes of the light nuclei. Thus, although we still place a high priority on the development of an interaction which can be used throughout the periodic table, it is clear that the more phenomenological potentials

TABLE V.

Wave functions for the oblate and prolate solutions of Si^{28}

Si^{28} (Oblate) $b_x = b_y = 1.877$ $b_z = 1.514$

Si^{28} (Prolate) $b_x = b_y = 1.684$ $b_z = 2.017$

Even Parity States

	Oblate				Prolate			
$(2k)^{\Pi}$	1^+	5^+	1^+	3^+	1^+	1^+	3^+	1^+
$n_x n_y n_z \sigma$ / ε	-63.0	-17.4	-15.2	-14.6	-59.7	-19.9	-15.0	-10.4
000	.999	.000	.036	.000	.999	.004	.000	.011
i110	.000	.707	.000	-.685	.000	.000	.074	.000
200	-.027	.500	.695	.485	-.035	-.014	-.052	-.080
020	-.027	-.500	.695	-.485	-.035	-.014	.052	-.080
002	-.028	.000	-.092	.000	.000	.955	.000	-.030
-101	.005	.000	-.110	-.174	-.012	.209	.703	.671
i011	-.005	.000	.110	-.174	.012	-.209	.703	-.671

Odd Parity States

	Oblate			Prolate		
$(2k)^{\Pi}$	3^-	1^-	1^-	1^-	3^-	1^-
$n_x n_y n_z \sigma$ / ε	-38.3	-36.4	-23.3	-39.8	-30.3	-26.0
001	.000	.157	.988	.994	.000	-.109
-100	.707	.698	-.111	.077	.707	.703
i010	.707	-.698	.111	-.077	.707	-.703

developed for a specific region of the table, here specifically the 2s-1d shell nuclei, will yield completely consistent results in that region. In particular, the effective one-body spin-orbit interaction appears to be an adequate approximation for the calculation of light nuclei.

We have further shown that a basis of Cartesian oscillator functions is extremely efficient in minimizing the effects of truncation. Although this advantage will only become more pronounced when one attempts to calculate the large deformations in the heavy fissionable nuclei, it must unfortunately be weighted against the huge number of two-body matrix elements which is required in this representation.

REFERENCES

1. W.H. Bassichis and J.P. Svenne, Phys. Rev. Lett., 18, 80 (1967); M.K. Pal and A.P. Stamp, Phys. Rev. 158, 924 (1967).

2. C.W. Nestor, K.T.R. Davies, S.J. Krieger, and M. Baranger, Nucl. Phys. A113, 14 (1968). The interaction used in the calculations is potential no. 1 of the above reference.

3. R.M. Tarbutton and K.T.R. Davies, Nucl. Phys. A120, 1 (1968).

4. G. Ripka, *Advances in Nuclear Physics,* Plenum Press, Inc., New York, 1968, Vol. 1.

5. A.P. Stamp, Nucl. Phys. A105, 627 (1967).

6. M.K. Banerjee, C.A. Levinson and G.J. Stephenson, Jr., Phys. Rev. 178, 1709 (1969).

7. R. Hofstadter and H.R. Collard, *Landolt-Bornstein,* Springer-Verlag, Berlin, 1967.

8. S.J. Krieger, Phys. Lett. 22, 97 (1969).

9. G.J. Wagner, Bull. Am. Phys. Soc. 14, 85 (1969); L.R.B. Elton and A. Swift, Nucl. Phys. A94, 52 (1967).

DISCUSSION

KUO: I would like to ask you about the nucleon-nucleon potential you used. Your potential approximately fits the nucleon-nucleon phase shifts. But when one works within a small model space, the model space effective interaction is very different from the free nucleon-nucleon interaction. The latter fits the phase shifts, and hence the former should not. So I feel that you could relax the requirement that your nucleon-nucleon potential should fit the phase shifts.

KRIEGER: That goes back to the two diverse ways of doing things in finite nuclei. The first way is to find the potential for which the Hartree-Fock approximation is suitable. Now in that case, the Hartree-Fock interaction and G-matrix are essentially identical. The interaction is weak enough, then the G-matrix is arbitrarily close to the potential and that, of course, is what we strive for. As I mentioned, it is empirical evidence that one can't make the potential that soft. On the other hand, the interaction in Hartree-Fock, the model state is a hole state because you don't have to worry about correlations so the interaction that one uses in Hartree-Fock is just the interaction which fits the phase shifts just as if one were to do a G-matrix calculation in which the model states was the entire space. The G-matrix would fit the phase shifts, because the on-shell G-matrix elements are essentially the phase shifts. So if one used the entire model state, one would want the G-matrix to fit the phase shifts.

KUO: When we do a usual Hartree-Fock calculation we can only work in a small space.

KRIEGER: Yes, the effects are, for example, the results that I showed you used only the orbital of the 2s-1d shell. If I include the 2p-1f levels, the energies change by less then one percent. That is, the interaction is smooth enough so that there are no further effects brought in for using the other, the higher lying states. They don't influence the results. That would not be true if the potential brought in very short range correlations. There you would have to have contributions from the high lying states. In the case which I am illustrating those correlations are very weak and it doesn't matter.

KUO: In short, I feel that you can have more freedom in the choice of your force. For example, we may use the effective forces derived by Siemen or Negele. They derived these forces from the Reid nucleon-nucleon potential, and their forces are designed to be used within a small model space. Their forces are very different from the free nucleon-nucleon force, and hence will not reproduce the phase shifts.

KRIEGER: Let me point out, for example, when one solves for the wave function for the scattering solution one finds there is no wound. There are two diverse ways of approaching the problem. This is a very, very soft potential. The place where you lose out is, as I mentioned, the binding energies. That's the price one pays for not working with a more realistic potential, which has more short range correlation.

ZAMICK: Have you performed these famous four-particle, four-hole calculations with this particular potential, and if so, where do the four-particle, four-hole states lie?

KRIEGER: I just calculated 0^{16} which was enough. If I remember correctly, the four-particle, four-hole state in 0^{16} comes out on the order of 20 MeV as one projects, if one projects by taking J/21 that tells you the center of gravity, it turns out on the experimental spectra--it comes down to something like 20, which is much too high. Since one is computing an energy difference rather than just an absolute energy, I think that that indicates that the higher correlations are important and must be included. So although I believe that this 6.06 state in 0^{16} is predominantly a four-particle, four-hole state, I would certainly say that it has significant admixtures of higher configurations.

ZAMICK: Did you get good quadrupole moments, i.e., effective charges with your Hartree-Fock calculations?

KRIEGER: They agree very well with the shell model results. I think that they are something like two-thirds too small.

HARVEY: In the comparison of quadrupole moment (Q) from Hartree-Fock shell model, one must remember that the shell model has the arbitrary effective charge parameter of 1/2. Presumably for an accurate comparison one should use an effective charge corresponding to the polarization of the core as seen in the Hartree-Fock calculation. It is my feeling that the Q from Hartree-Fock should be larger than the Q from shell model since the Hartree-Fock procedure maximizes the collective effects while the shell model can sample all types of correlations.

KRIEGER: The effective charge is an additional parameter. I was just comparing in a sense that they try to do their best to fit. I shouldn't say that the comparison is the best fit.

HARVEY: Concerning non-axially symmetric states--if χ is a non-axially symmetric state, then the expansion of χ in terms of states ψ_J of definite angular momentum $(\chi = \Sigma a_J \psi)$ J will invove the states ψ_J corresponding to those in the Davydov-Fillipov model. One can, on the other hand, first expand the χ in terms of states of definite K$\chi = \psi i b_k \phi_k$, and then project states of definite J from the ϕ_K $\phi_K = \Sigma_i C_{JK} \psi_S$. Clearly, there is the relationship $\psi_J = \Sigma_K \frac{C_K C_{JK}}{a_j}{}_J$. The

The question now remains whether the state ψ_J or ψ_S or better approximately to the eigenfunctions of the Hamiltonian. In an old calculation by Elliott and myself in Mg^{24} we tested this by diagonalizing the Hamiltonian among the K=024 bands from the (84) representatives of SU_3 (this has an axially, symmetric determental intrinsic state). We found that K was a good quantum number, i.e., it is the $\tilde{\psi}_S$ that are the more appropriate functions and not the ψ_S. N.B. the intrinsic state ϕ_K of the $\tilde{\psi}_J$ are *not* determinants.

KRIEGER: One additional model which can be solved essentially exactly is the Lipkin model which I think Ripka worked on, and there the instability is associated, or the stability happens at a place the RPA breaks down. You can see that again, the instability occurs in a case which corresponds to something physical.

III. a. The Anatomy of the Hartree-Fock Procedure in a Simple Model

M. Harvey
Atomic Energy of Canada Limited
Chalk River Nuclear Laboratories
Chalk River, Ontario

It is difficult to go to any nuclear physics meeting these days without someone talking about Hartree-Fock - I am as guilty as anyone else. Until recently these calculations were confined to light-nuclei and the discussion of collective deformations. More recently these calculations have been taken to heavy nuclei and even super heavy nuclei in the discussion of binding energies and stabilities. My own interest has been to develop the self consistent theory for adiabatic fission.

The more we have worked at the self-consistent problem the more evident it was that we needed a complete analysis of a simple problem that would shed some light into the peculiarities that sometimes occur in seeking Hartree-Fock solutions. With this simple problem one can see under what conditions the Hartree-Fock solution is a good approximation to the lowest eigen solution to the Hamiltonian, and whether then all the Hartree-Fock solutions have any physical meaning. It is of interest also to study the Hartree-Fock iterative procedure. It is well known that iterative procedures do not always converge to a root of the problem - even when this root exists. Does this lack of convergence have any meaning or is it just a fault in the mathematics?

Let me briefly describe the problem to you and some of the analysis we have made - I clearly do not have time to go into any great detail.

We consider two fermions in a four dimensional space. The space is divided into a two dimensional orbital space and a two dimensional spin space. The Hamiltonian is assumed to be two body and act only in orbital space. If the orbital space is described by orbitals ϕ_1 and ϕ_2, then the two body matrix elements are represented by:

$$
\begin{array}{cccc} s=0 & s=0 & s=0 & s=1 \\ (\phi_1^2) & \overline{(\phi_1\phi_2)} & (\phi_2^2) & (\phi_1\phi_2) \end{array}
$$

$$
H = \begin{bmatrix} \alpha & \beta & \gamma & 0 \\ \beta & \delta & \varepsilon & 0 \\ \gamma & \varepsilon & \nu & 0 \\ 0 & 0 & 0 & \omega \end{bmatrix} \tag{1}
$$

The Hartree-Fock problem is to find the determinantal state $\chi = (\psi^2)^{s=0}$ which is lowest in energy with the orbital state ψ given by $\psi = m_1\phi_1 + m_2\phi_2$. Clearly because of the normalisation condition $m_1^2 + m_2^2 = 1$, this is a one parameter problem in $r = m_1/m_2$ which is why we chose it.

It is of course easy to write down the total energy of χ as a function of r

$$E = (\chi|H|\chi) = (\alpha r^4 + 2\sqrt{2}\beta r^3 + 2(\delta+\gamma)r^2 + 2\sqrt{2}\,\varepsilon r + \nu)/(1+r^2)^2 \tag{2}$$

Clearly the extrema are defined by

$$\frac{\partial E}{\partial r} = 0 = -\sqrt{2}\beta r^4 + 2(\alpha-\delta-\gamma)r^3 + 3\sqrt{2}(\beta-\varepsilon)r^2 - 2(\nu-\delta-\gamma)r + \sqrt{2}\varepsilon \tag{3}$$

Note that, since the determinantal state χ is symmetric in orbital space, neither E nor $\frac{\partial E}{\partial r}$ can depend on the matrix element ω. For the discussion we shall choose matrix elements such that equation 3 has roots -2, $-1/44$, 2 and 4.

i.e. $\beta = -1/4\sqrt{2}$ $\qquad \varepsilon = +1/11\sqrt{2}$

$$4(\alpha-\delta-\gamma) = -175/88 \text{ and } 4(\times-\delta-\gamma) = -175/22 \tag{4}$$

One can now draw the curve for $(E-\delta-\gamma)$ as a function of r as shown in Figure 1. ($\delta+\gamma = c/4$ merely shifts the energy scale.) Clearly the roots $-1/44$ and 4 correspond to minima and -2 and 2 correspond to maxima.

This, of course, is not the way that the Hartree-Fock procedure goes about finding extrema. In Hartree-Fock we set up the average single particle field as created by all but one particle of the system. The Hartree-Fock single particle Hamiltonian is thus a function of the structure of the single particle orbits that it creates; i.e. there is self consistency. In our problem, the single particle Hamiltonian h does not have any spin dependent terms and thus has a representation only in the two dimensional orbital space.

$$\begin{bmatrix} h_{11} & h_{12} \\ h_{21} & h_{22} \end{bmatrix} = \varepsilon \begin{bmatrix} m_1 \\ m_2 \end{bmatrix} \tag{5}$$

$$\begin{aligned} h_{11} &= \alpha m_1^2 + \sqrt{2}\beta m_1 m_2 + \tfrac{1}{2}(3\omega+\delta)m_2^2 \\ h_{21} = h_{12} &= \sqrt{\tfrac{1}{2}}\beta m_1^2 - \tfrac{1}{2}(3\omega-\gamma)m_1 m_2 + \sqrt{\tfrac{1}{2}}\varepsilon m_1^2 \\ h_{22} &= \tfrac{1}{2}(3\omega+\delta)m_1^2 + \sqrt{2}\varepsilon m_1 m_2 + \nu m_2^2 \end{aligned} \tag{6}$$

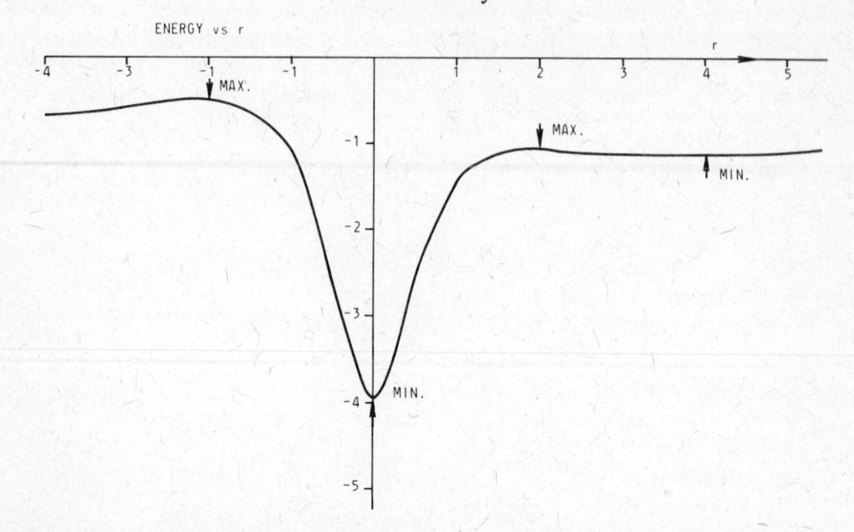

Fig. 1. Energy (E) of the determinantal state $\chi = (\psi^2)^{S=0}$ plotted against the parameter $r = m_1/m_2$ for a Hamiltonian of Eq. (1) under conditions in Eq. (4). [Actually we have shifted the energy zero by plotting $(E-\delta-\gamma)$].

The diagonalization of the matrix h together with the first of equation 5 leads to the condition

$$(h_{22}-h_{11})r + h_{12}(r^2-1) = 0 \tag{7}$$

On substituting for the h one can verify that equation 7 is exactly equation 3 (as it should be). Note that the matrix elements of h are functions of ω but, on substitution in 7, the coefficient of ω vanishes.

The iterative procedure in Hartree-Fock treats equation 7 as a quadratic equation in r with constant coefficients h_{ij}.

$$r = \frac{(h_{11}-h_{22}) \pm \quad (h_{11}-h_{22})^2 + h_{12}^2}{2h_{12}} = G^{\pm}$$

$$\varepsilon = h_{21}r + h_{22} = \tfrac{1}{2}(h_{11}+h_{22}) \pm \quad (h_{11}-h_{22})^2 + 4h_{12}^2$$

In the Hartree-Fock procedure we usually seek the lowest root and so iterate on the system.

$$f = G^{-}(r)$$

Note that the function G^{-} through the h_{ij} is now a function of ω. Thus the antisymmetric matrix element ω can affect the iterative procedure even though the roots of the equation are independent of ω. To illustrate the iterative procedure we draw the curves $y = r$ and $y = G^{-}(r)$; where these cross (at r_0) is a root to the problem (an extrema). If $-1 < \frac{\partial G}{\partial r}\Big|_{r_0} < 1$, then the iterative procedure will converge to the root otherwise it will diverge or flip-flop. This is illustrated in Figure 2. The actual function G depends on ω through the quantity $f-\frac{1}{2}C = 3\omega-\delta-2\gamma$. We show the curves $y = G(r)$ for various values of $f-\frac{1}{2}C$ on Figure 3. It can be shown that for $f-\frac{1}{2}C > .4545$ the function $y = G(r)$ cuts $y = r$ at all the four roots including those corresponding to the maxima at $r_0 = \pm 2$. Note however that the minimum solutions at $- 1/44$ and 4 are stable in the iteration procedure whereas the maxima are unstable. As the quantity $f-\frac{1}{2}C$ decreases past .4545, the curve for G(r) changes dramatically to miss the larger maximum root at $r_0 = -2$. Further decrease of $f-\frac{1}{2}C$ past $-.4545$ sees the root at $r_0 = -2$ missed; then for $f-\frac{1}{2}C < -.9773$ the minimum root at $r_0 = 4$ is missed and finally for $f-\frac{1}{2}C < -3.995$ none of the roots, including that at the absolute minimum satisfy the iteration procedure. This dependence of the solution to the iterative problem depending on a quantity $(f-\frac{1}{2}C)$ for which the roots themselves are not functions may seem from the point of view of the mathematics to be very strange. I think we can give the Hartree-Fock iterative procedure a little more physical meaning however and not treat it as simply a means to finding roots to complicated equations. In this simple problem the decrease of the quantity $f-\frac{1}{2}C$ implies the lowering in energy of the space-antisymmetric state $(\phi_1\gamma\phi_2)^{s=1}$. The iterative procedure thus feels the presence of this state and, when it is too low in energy warns us (by failing to converge) that an attempt at describing the lowest solution by a space-symmetric function χ is very poor.

There are many facets to this problem that I will not dwell on here. I merely want to make a remark finally about energy surfaces.

I have shown earlier in Figure 1 the energy given as a function of r. This is not the normal way we represent surfaces in nuclear physics. Usually we draw the energy as a function of the expectation value of a single particle operator Q; often Q is the quadrupole operator but this is unimportant for the present arguments. Every single particle operator is proportional to the density matrix ρ and the energy from a two-body operator is proportional to ρ^2.

The matrix elements of ρ are always quadratic forms in the expansion coefficients $m_{\alpha i}$ of Hartree-Fock orbital α in terms of the basis set i $(\rho_{ji} = \sum_{\alpha<} m_{\alpha i}\, m_{\alpha j})$. Thus in drawing energy surfaces we are trying to draw a function of the fourth power in expansion coefficients $m_{\alpha i}$ again the function $\langle Q\rangle$ which is of the second power. Not surprisingly we can get very strange looking results. In Figure 4 I show the energy curve of Figure 1 again and also the form it takes

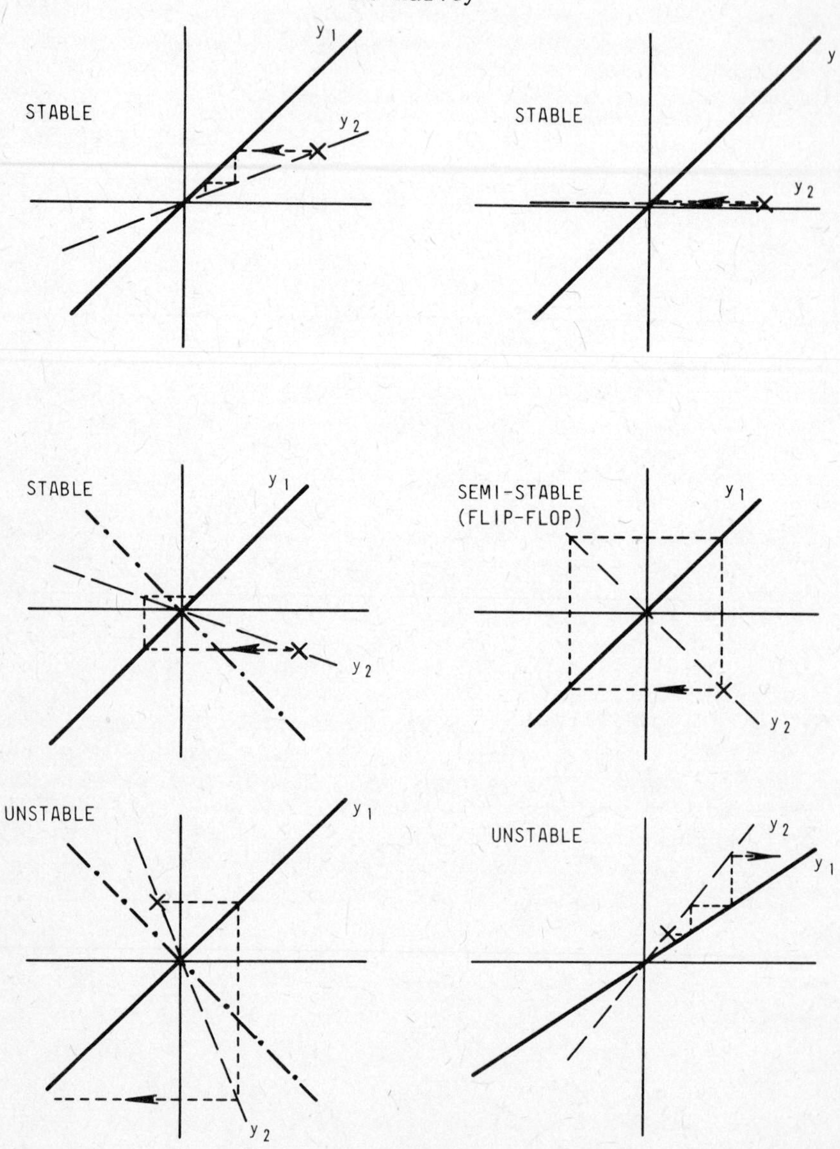

Fig. 2. Illustrations demonstrating the convergence of an iterative procedure $r = G(r)$. The thick, solid line shows the curve $y = y_1 = r$. The long-dashed line y_2 shows the slope $G(r)$ where it cuts the line $y = r$. The arrowed dashed line indicates the direction of iteration when started at a point near to a root (i.e. near r_0 where $r_0 = G(r_0)$): the various diagrams have been labelled according to whether the iteration procedure does or does not converge to the root. In general

$$\text{convergence occurs when } \left|\frac{dG}{dr}\right|_{r_o} < 1.$$

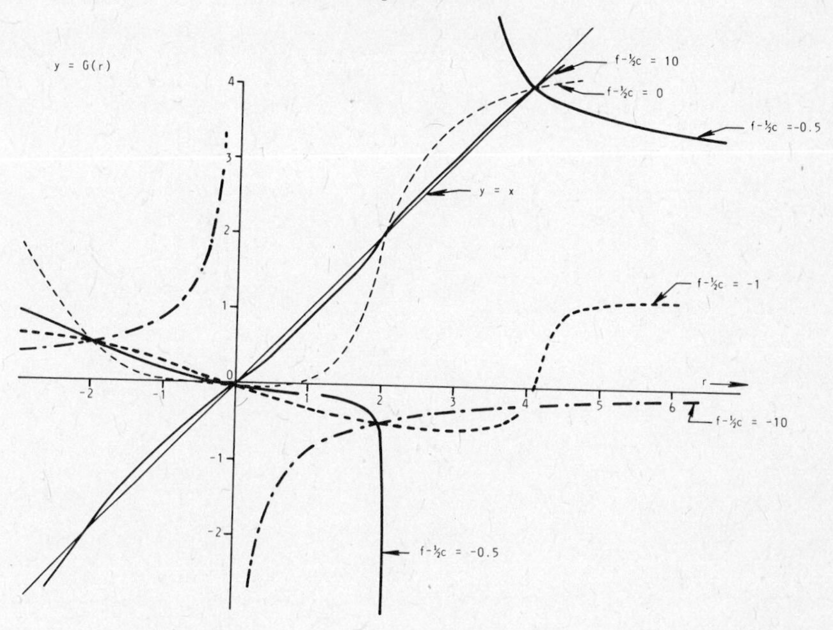

Fig. 3. Plots of $y = G(r)$ for various values of the parameter $f - 1/2\ C = 3\omega - \delta - 2\gamma$. Roots occur where $y = G(r)$ cuts the line $y = r$ and the stability of the solution can be deduced by comparison with Figure 2. The curve $y = G(r)$ need not cut $y = r$ at a particular root (say r_0'): if it does not then it can be shown that $G(r_0') = -1/r_0'$.

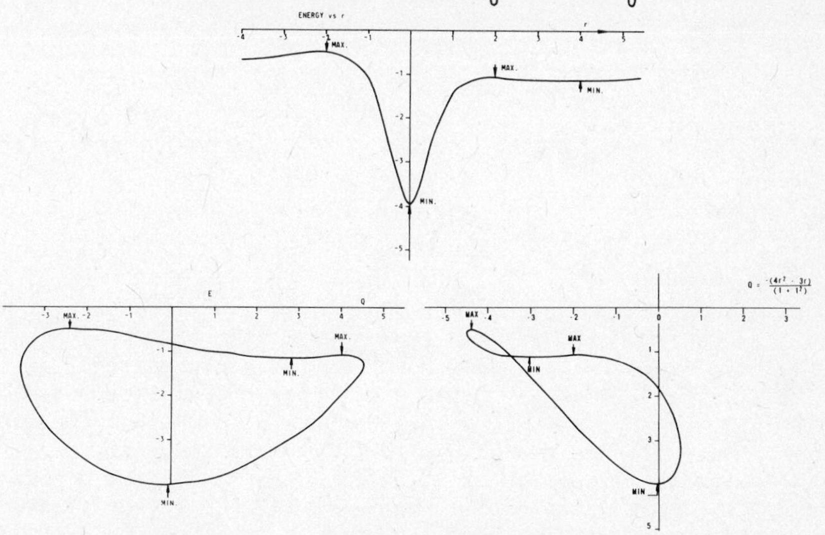

Fig. 4. The energy surface of the determinantal state χ plotted against r (as in Figure 1) compared with the energy plotted against the one body operators Q_1 (on the left) and Q_2 (on the right) with Q_i defined in the text.

when plotted against two different single particle operators $Q_i (i=1 \text{ and } 2)$

i	=	1	2
$(\phi_1\|Q_i\|\phi_2)$	=	½	−2
$(\phi_1\|Q_i\|\phi_2)$	=	0	0
$(\phi_2\|Q_i\|\phi_2)$	=	2	3/4

ACKNOWLEDGEMENTS

I wish to thank Axel Jensen for discussions on the Hartree-Fock problem.

III.B. ALPHA SCATTERING AND NUCLEAR STRUCTURE IN THE CALCIUM REGION

R. Santo
Sektion Physik der Universität, München
and
G. Gaul, H. Ludecke, H. Schmeing and R. Stock
Max-Planck-Institut für Kernphysik, Heidelberg

(Presented by R. Santo)

I. INTRODUCTION

Much information about models of nuclear scattering as well as the structure of excited states has been accumulated by studies of elastic and inelastic scattering of alpha-particles[1], but nearly all this information is obtained from analyses of angular distributions up to about $\Theta = 70°$. Previous measurements of the $Ca^{40}(\alpha,\alpha)$ Ca^{40} (Ref. 2-4) and K^{39} (α,α) K^{39} (Ref. 5) scattering have revealed a strong increase of the cross section at large angles accompanied by an irregular energy dependence. On the other hand, data in the Titanium-Nickel region[6,7] showed regularly decreasing angular distributions up to about 170°. This rapid change in the character of the angular distributions is not expected from the simple optical model, according to which all relevant quantities should vary smoothly with energy or mass number. The experiments to be discussed in the following have been done in order to get more insight into the character of the "backward anomalies" and their connection with nuclear structure.

II. EXPERIMENTAL PROCEDURE

The experiments have been performed at the Heidelberg MP-Tandem using a 90 cm scattering chamber and eleven surface barrier counters mounted on the movable lid of the chamber. For the Calcium targets isotopically enriched $CaCO_3$ has been evaporated onto thin carbon backings. For the Argon measurements a gas target has been used in connection with a special collimating system in front of the detectors. The resolution in the $Ca(\alpha,\alpha)$ spectra was about 60 keV. In most cases, angular distributions have been measured in 2° steps at different energies.

A. Elastic Scattering*

At angles smaller than about 70°, the angular distribu-

* Parts of these data have been published in Ref. 8.

tions from Ca^{40}, Ca^{42}, Ca^{44} and Ca^{48} all are very similar, at least at energies where comparisons have been made[8,9]. It is clear that these parts of the angular distributions are well described by a smooth optical model potential - for instance, the average potential found in the analysis of data in the Ti-Ni region 6. (In the following the name "average" potential will be used exclusively for the potential V = 183.7 MeV, W = 26.6 MeV, r_0 = 1.4 f, a = 0.564 f of ref. 6). Looking at the entire angular distributions up to 180° one recognizes large differences between the single isotopes. Thereby, the doubly magic Ca^{40} is anomalous, whereas the Ca^{44} curve fits well into the systematics for heavier nuclei and, consequently, is fitted by the average potential. Figure 1 illustrates this behavior.

It should be noted that in the available data for the scattering of deuterons or He^3-particles on the Ca isotopes[10,11] no similar deviations have been observed.

To establish a systematics of the alpha-scattering anomalies we performed a comparative experiment on the Argon isotopes. The results are shown in Figure 2. Again a backward rise is observed, now for Ar^{36}. The comparison of the experimental data (see Figure 3) shows the following:

(i) There is a great similarity between the N = Z nuclei Ca^{40} and Ar^{36} on one hand, and the N = Z + 4 nuclei Ca^{44} and Ar^{40} on the other hand.

(ii) The N = Z group shows the backward rise, whereas the N = Z + 4 group behaves regularly.

(iii) The previously observed close correspondence between the angular distributions from Ca^{40} and K^{39} (Ref. 5) extends to their N + 2 isotopes, which at least at two measured energies are very similar (Ref. 8).

From excitation function measurements of the Ca^{40} (α,α) reaction[12] we conclude that compound nuclear contributions are negligible even at large angles. On the other hand, it is clear that a smooth average optical model potential can not at the same time describe the regular (Ca^{44}, Ar^{40}) angular distributions and the anomalous (Ca^{40}, Ar^{36}) data with their irregular energy dependence. Modifications of the optical model or additional processes have to be considered to explain this large difference between neighboring nuclei. Our starting point will be the total real potential- nuclear + Coulomb + centrifugal - seen by a particular partial wave at a certain incident energy. For a large class of optical potentials (e.g. the average potential) this total potential exhibits a relative minimum followed by a secondary maximum near the nuclear surface. Partial wave resonances can lead to backward rises in the angular distributions[2] and to an irregular energy dependence. For a realistic description of alpha-particle scattering, however, we have also to consider the absorption. Regularly decreasing angular distributions and a smooth energy behavior may indicate that

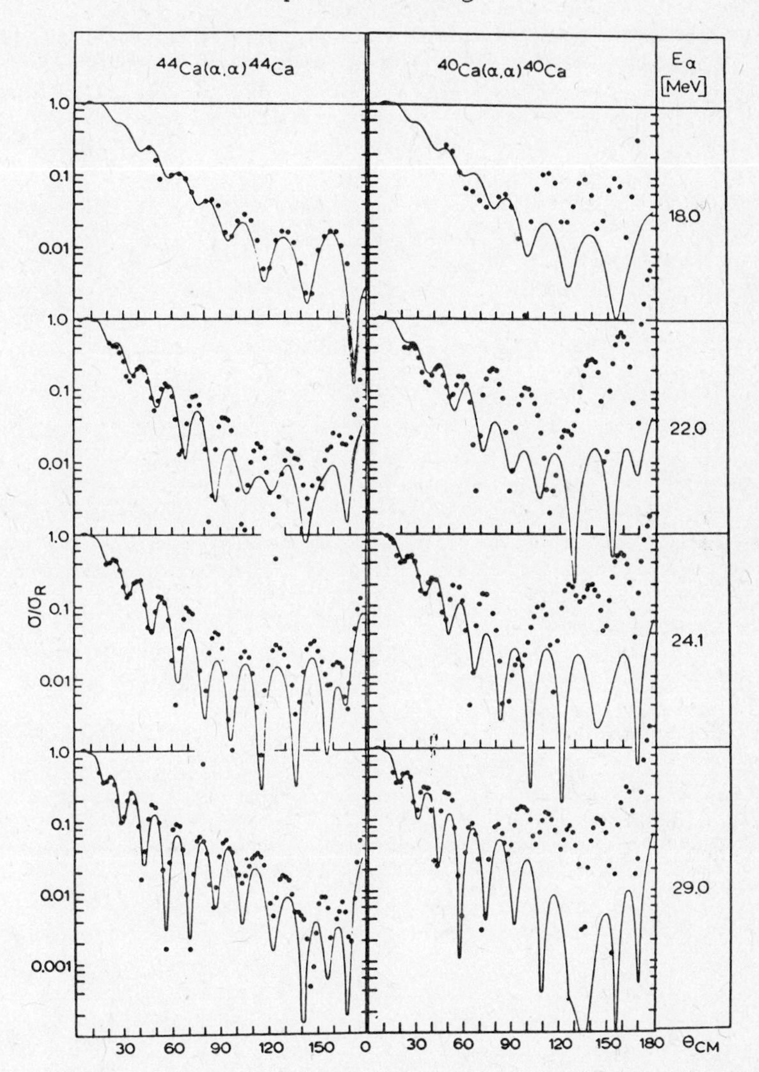

Fig. 1. Elastic alpha-scattering angular distributions. The solid lines represent optical model calculations using the average parameters as described in the text.

the absorption is sufficiently strong to prevent particles from feeling the details of the real potential inside the nucleus. If the imaginary potential is considerably reduced in the surface region (as compared, for instance, to the average potential parameters) the cross section at backward angles is increased. In this way, by arbitrarily varying the (six) parameters of an optical model potential, the anomalous Ca^{40} and K^{39} data have been fitted, but rather different parameters were required to fit the individual

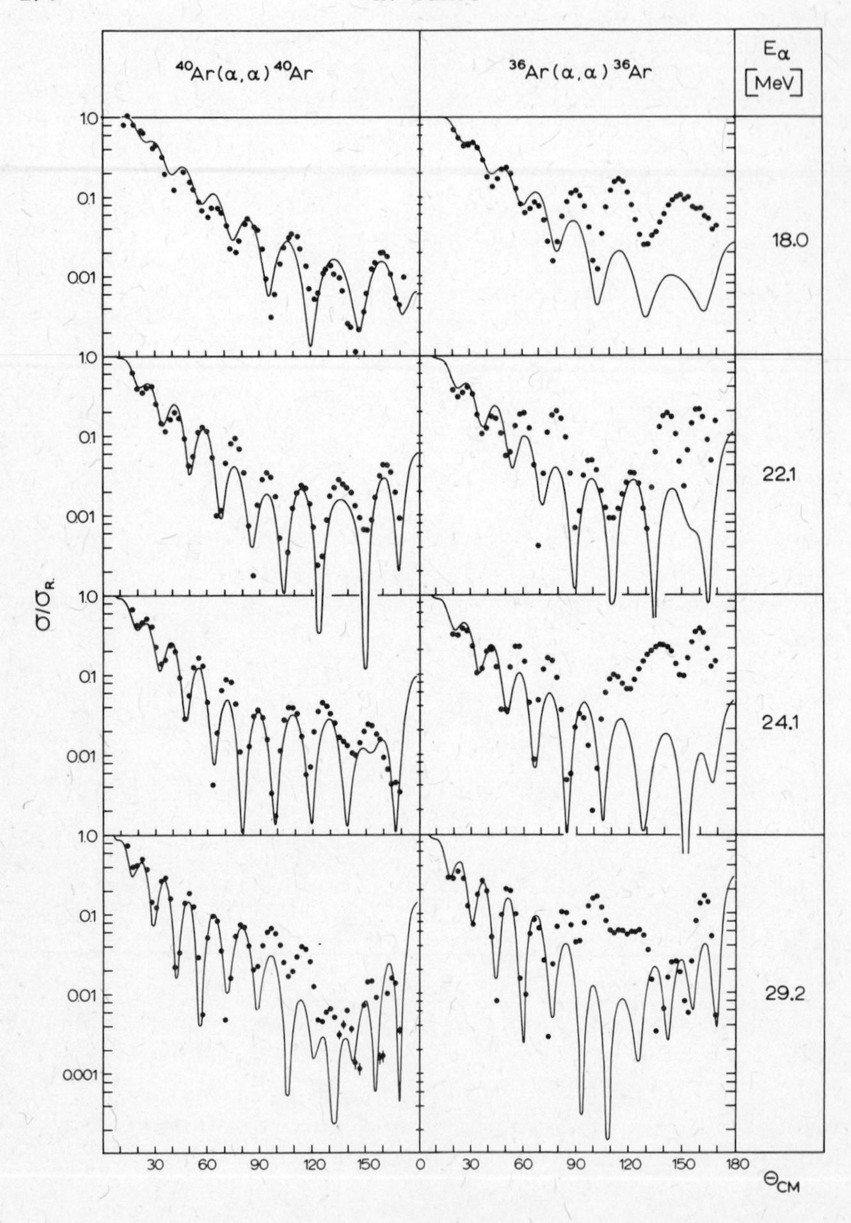

Fig. 2. Elastic alpha-scattering angular distributions. The solid lines represent optical model calculations using the average parameters as described in the text.

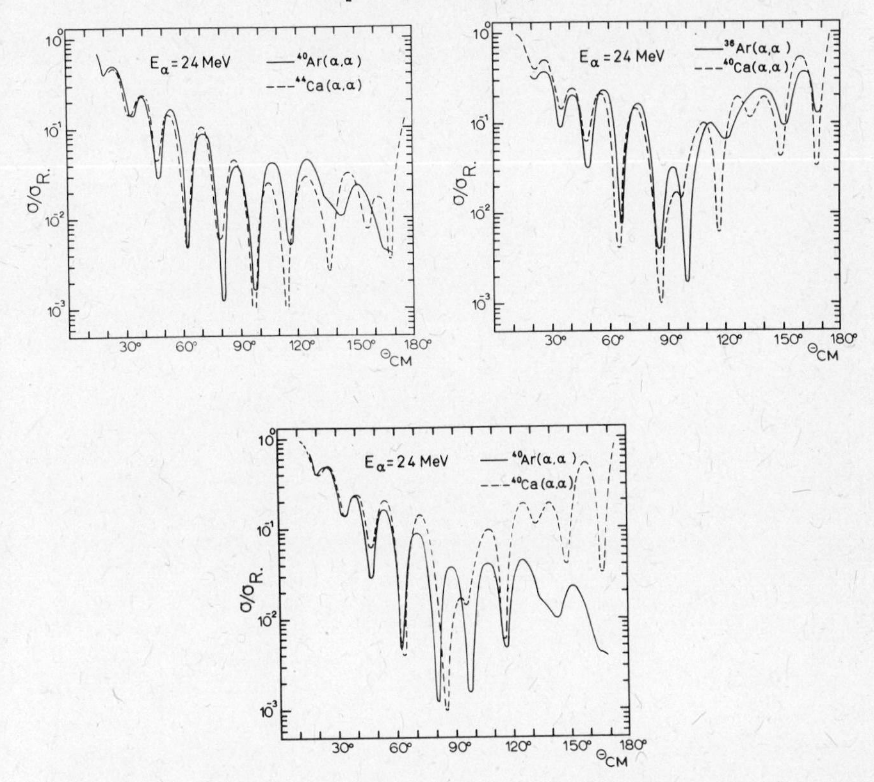

Fig. 3. Comparison of experimental angular distributions.

energies. Furthermore, the fit to the back angles is payed by a worse fit at forward angles. The results of these optical model searches indicate that the observed anomalies lie outside the domain of the simple optical model.

A modification of the optical model which leads to a partial reduction of the absorption has been proposed by Chatwin et al.[13]. In this model absorption only takes place for partial waves smaller than a critical angular momentum L_c. This L_c is determined by comparing the angular momenta carried in by the elastic channel with the angular momenta carried away by the various reaction channels. Good fits to Ca^{40} (α,α) data for $E \leqq 18$ MeV have been obtained by Bisson and Davis[14] using L_c as a fitting parameter and an optical potential which already fitted Ca^{40} data between 5 MeV and 12 MeV. Attempts are now being made at Tallahassee[15] to connect L_c with angular momenta of the various compound nuclear reaction channels. Using the average potential and varying L_c we did not succeed in obtaining a reasonable fit to the Ca^{40} (α,α) data between 20 MeV and 29 MeV. Now the influence of using different optical model potentials has to be studied.

Similar restrictions on angular momenta as considered by Chatwin *et al.* may result from direct reaction channels strongly coupled to the elastic entrance channel. In the case of Oxygen-Oxygen scattering - which bears much similarity with our alpha-scattering - such couplings have been considered by Shaw *et al.*[16]

B. Effects of Alpha-Correlations

In the simple optical model, antisymmetrization between the incident alpha-particles and the target nucleons is not taken into account properly. If the target nucleus contains appreciable amounts of alpha-correlations, characteristic deviations from the simple optical model behavior are expected. If we wish to preserve the common optical model these effects have to be formulated as separate processes and added to optical model scattering. The additional amplitudes depend on the amount of alpha-correlation in the states involved, leading to a natural explanation for the isotopic dependence of the anomalies: alpha-clusters are preferentially formed in N = Z nuclei and are rapidly destroyed by excess neutrons. A very strong blocking effect of this type has been found by Groß and Eichler[17] in a schematic model calculation of alpha-correlations.

So far, scattering processes involving alpha-clusters have only been calculated for simple cases and under rather restrictive assumptions[18-20].

Instead, one may handle these additional interactions as phenomenological potentials added to the average potential. Guided by the idea that alpha-clusters are most likely at the nuclear surface we added to the average potential a Gaussian potential centered near the surface. It turned out that with fixed values of this additional potential the entire angular distributions of the $Ca^{40}(\alpha,\alpha)$ scattering at our four energies (18, 22, 24, and 29 MeV) are rather well described.

C. Inelastic Scattering

More insight into the anomalies of elastic scattering may be gained by looking at the inelastic scattering. Therefore we measured the inelastic alpha-scattering on Ca^{40} and Ca^{44} at 29 MeV up to 176°. (A careful investigation of alpha-scattering on Calcium isotopes comprising angular distributions up to about 70° has been made by Lippincott and Bernstein[9] at 31 MeV with about 100 keV resolution.)

Some of our angular distributions are shown in Figure 4. It is seen that the data may be classified according to their behavior at large angles. First, there are angular distributions which decrease monotonically towards larger angles. Of these, the cross section for the 2^+ state has the most

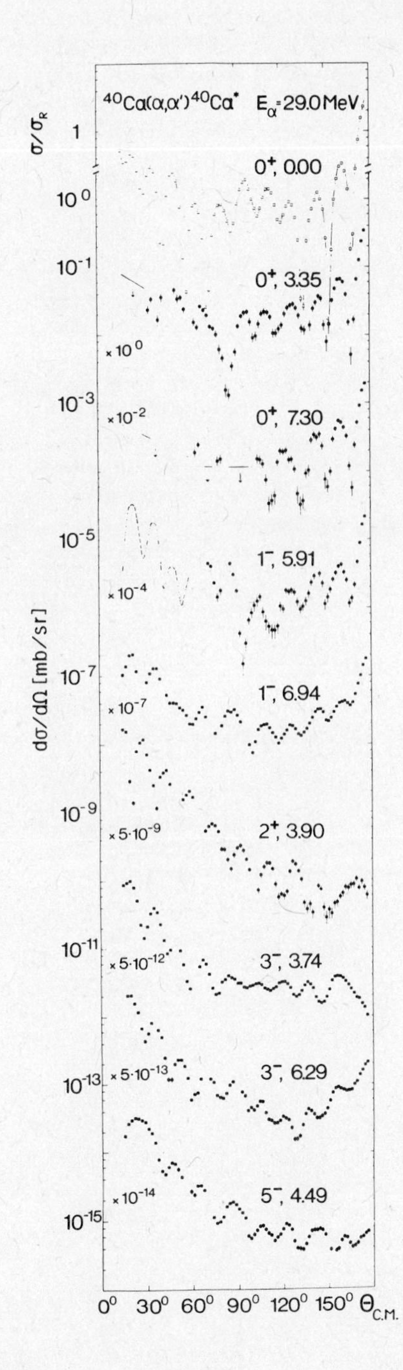

Fig. 4. Experimental cross sections for the inelastic scattering of 29 MeV alpha particles on Ca^{40} and the ratio to Rutherford scattering for the elastic scattering. Spins and excitation energies are indicated.

typical diffraction shape for a first order direct interaction over the entire angular range measured. Similarly, the 5^- state and the 3^- state at 3.74 MeV have diffractional, decreasing angular distributions largely obeying the Blair phase rule. In the $Ca^{44}(\alpha,\alpha)$ experiment *all* measured angular distributions - including the first 0^+ state - display this smooth behavior. For the states in the second and unusual class, the cross section becomes larger at back angles the envelope increasing at 120° and ending with a large peak at 180°, just as for the groundstate. The 1^- and 0^+ states clearly belong to this class.

It should be noted that, for the first time, excited 0^+ states have been observed in inelastic alpha-scattering on Ca^{40}. The 0^+ states at 3.35 MeV and 7.30 MeV show a pronounced structure and rather large cross sections at back angles disappearing into the background at forward angles. Apart from our interest in the mechanism of the anomalies the back angle behavior of the 0^+ states may be important as a method for finding such states.

Near 180° the 1^- state at 6.94 MeV is an order of magnitude more strongly excited than any other state. Its cross section is about 200 times larger than that of the 2^+ state. By contrast, the 2^+, 3^- and 5^- states, probably collective, dominate the spectra at forward angles.

From the discussion of elastic scattering and the optical model analysis it is clear that similar consistency problems arise in applying the standard DWBA to the inelastic angular distributions of Figure 4. If we confine ourselves to $\theta \leqq 70°$ we can get good fits to the Ca^{40} and Ca^{44} data by using the average parameters and a collective model formfactor. Furthermore, the 2^+ state is well described by this standard DWBA over the entire angular range, and similarly for the 5^- state. The angular distributions in the backward rise class, however, exceed by far DWBA predictions. Also, their structure is still pronounced at excitation energies of 7-8 MeV, where standard DWBA calculations show a much smoother structure due to the energy difference between the entrance and exit channels.

There seems to be no unique correlation between the final spin, excitation energy, or excitation strength and the back angle behavior. The 3^- state at 6.29 MeV, for instance, seems to have a backward peaked cross section.

The selectivity of the excitation mechanism at back angles, which has been discussed above, indicates that this effect is not a mere reflection of the elastic scattering anomalies. On the contrary, the microscopic structure of the excited states and the specific way in which they are excited in alpha-scattering must be responsible for the behavior at large angles.

REFERENCES

1. A. M. Bernstein, Advances in Nuclear Physics 3 325 (1969) An exhaustive list of papers concerning alpha-scattering is found in this reference.

2. C. R. Gruhn and N. S. Wall, Nucl. Phys. 81 161 (1966).

3. A. Budzanowski, K. Grotowski, L. Jarczyk, Mrs. B. Lazarska, S. Micek; H. Niewodniczanski, A. Strazalkowski and Mrs. Z. Wrobel, Phys. Lett. 16 135 (1965).

4. E. T. Boschitz, J. S. Vincent, R. W. Bercaw and J. R. Priest, Phys. Rev. Lett. 13 442 (1964).

5. A. Bobrowska, A. Budzanowski, K. Grotowski, L. Jarczyk, S. Micek, H. Niewodniczanski, A. StrazaLkowski and Z. Wrobel, Nucl. Phys. A126 361 (1969), A126 369 (1969).

6. R. Bock, P. David, H. H. Duhm, H. Hefele, U. Lynen and R. Stock, Nucl. Phys. A92 539 (1967).

7. L. McFadden and G. R. Satchler, Nucl. Phys. 84 177 (1965).

8. G. Gaul, H. Lüdecke, R. Santo, H. Schmeing and R. Stock, Nucl. Phys. A137 177 (1969).

9. E. D. Lippincott and A. M. Bernstein, Phys. Rev. 163 1170 (1967).

10. W. Fitz, J. Heger, R. Santo and S. Wenneis, Nucl. Phys. A143 113 (1970).

11. H. P. Morsch, private communication.

12. H. Schmeing *et al.*, to be published.

13. R. A. Chatwin, J. S. Eck, D. Robson and A. Richter, Phys. Rev. C1 795 (1970).

14. H. E. Bisson and R. H. Davis, Phys. Rev. Lett. 22 542 (1969).

15. K. Eberhard, private communication.

16. R. W. Shaw, Jr., R. Vandenbosch and M. K. Mehta, Phys. Rev. Lett. 25 457 (1970).

17. P. H. E. Gross and I. Eichler, to be published.

18. T. Honda, Nucl. Phys. A136 183 (1969).

19. N. C. Schmeing, Nucl. Phys. A142 449 (1970).

20. W. R. Thompson, to be published.

DISCUSSION

ENDT: Does the resonance observed in alpha-particle scattering on Ca^{40} have any bearing on the back-angle effects you observe?

SANTO: One has to be careful to speak of resonances since even the optical model calculation gives you some kind of a resonance in the 180° cross section.

ENDT: I remember from the Krakow data that the resonance was rather narrower and also rather higher than you are showing now.

SANTO: Yes, that's right, there is, apart from the difference in the absolute cross sections, an enhancement compared to the average optical model prediction, but it is difficult to see what is resonance and what is optical model behavior. We measured these data and we don't find an effect of the same magnitude.

ENDT: My other question was: what about alpha-particle backward scattering of let's say S^{32} or maybe Si^{28}?

SANTO: We did this quite recently and S^{32} again shows the same anomalous behavior, but now we are trying to compare with S^{34}.

The effect may be different in the s,d-shell compared to the $f_{7/2}$ shell. I think that structural calculations show that pairing is not so important in the s,d shell, so that alpha correlations may be favored and, on the other hand, the pairing becomes important in the $f_{7/2}$ shell, so pairing is competing with alpha correlations if one goes from the s,d-shell to the $f_{7/2}$ shell.

KRIEGER: There is a big difference between the pairing and deuterons and there is a big difference between four n nuclei and alpha particles. I mean, the nuclei we are discussing are all four n nuclei and so they have the correct spin-isospin to form alpha particles, but I have to admit that I would be amazed if that they in fact did it. There is spacial correlation that one requires and I think that the reason they are called alpha particles nuclei is because a theorist can easily form a particle model of them, but whether that model of them is really found in nature I think is something else again. In the same way that Ca^{44} could be an alpha particle nucleus with the four extra neutrons being exchanged among these alpha particles, so I don't know -- it's a nice model but to find out if it has any basis in fact is difficult.

SANTO: But a simple alpha-particle model, for instance, gives you a much better O^{16} binding energy than most other nuclear model calculations do.

KRIEGER: No, I would say about the same thing on Ca^{40}.

IV.A. COLLECTIVE CALCULATIONS IN LIGHT AND INTERMEDIATE MASS NUCLEI

Larry Zamick*
Department of Physics
Rutgers - The State University
New Brunswick, New Jersey

I. INTRODUCTION

I have been asked to talk mainly about particles coupled to vibrations. Consider as an example the surprisingly low lying $J = 3/2^+$ state in Ca^{41} at 2.02 MeV whose parity is opposite to that of the $7/2^-$ ground state. We can describe this state as an $f_{7/2}$ nucleon coupled to the lowest 3^- and 5^- vibrational states of Ca^{40}. This is an example where particle *X VIBRATION* forms the majority of the wave function of a given state.

We now come to examples where particle *X VIBRATION* forms a *small* but *important* part of a wave function. The ground state of Ca^{41} would have no charge quadrupole moment if the shell model assignment, $f_{7/2}$ Neutron $| Ca^{40} >$ were literally true. However a small admixture of the configuration $f_{7/2}$ x 2^+ *VIBRATION* of Ca^{40} can account for its quadrupole moment (which has not yet been measured, come to think of it; but which undoubtedly would be approximately the same as if the $f_{7/2}$ neutron were assumed to be a proton). The admixture is small enough so as not to be noticed, at least not with any conviction, in one nucleon transfer reactions i.e. spectroscopic factors. The above admixture provides justification for the intuitive idea of effective charge, as was shown by Arima and Horie[1] and by Fallieros and Ferrel.[2]

The coupling of $f_{7/2}$ neutrons to the monopole (0^+) vibrations of Ca^{40} might account for the fact that these isotopes do not all have the same charge radius. But strange things are happening in this case, as we shall see later.

The measured magnetic moment of Co^{55} ($f_{7/2}^{-1}$ proton x Ni^{56}) is $4.2 \rightarrow 4.5$ nm., considerably smaller than the Schmidt value 5.79 nm. A good part of this comes from a small admixture into the ground state wave function of $f_{7/2}^{-1}$ x $(J = 1^+ \ T = 1)$ vibration of Ni^{56}.

II. BREAK DOWN OF THE INTUITIVE IDEA OF EFFECTIVE CHARGE

Intuitively, we think that if one nucleon has a given effective charge, then when we go to two nucleons in a given shell we can assign each of them the same effective charge as one nucleon. (I use effective charge in the most general

* Supported in part by The National Science Foundation.

sense to include effective magnetic g factors etc.).

Let me give you a trivial example where this is not so. Let our model space be $f_{7/2}{}^n$ in the Calcium isotopes. We know that as a result of the admixture of the configuration $f_{7/2}{}^{n-1}\ f_{5/2}$ the magnetic moment gets quenched. But consider $n = 1$ i.e. Ca^{41}. I cannot admix to my basic $f_{7/2}$ wave function the configuration $f_{5/2}$ because this violates angular momentum conservation. Hence there is no one-body renormalization of the effective charge. Consider next two neutrons in a $J = 6$ state, we can admix $(f_{7/2}\ f_{5/2})^{J = 6}$ to the basic $(f^2_{7/2})^{J = 6}$ configuration and therefore account for the fact that the g factor for this state is smaller than it is for Ca^{41}. Remember that in the pure $f_{7/2}$ approximation $\mu = g\,J$ for *all* states where g is *fixed* independent of n or J ($g = \frac{-1.91}{3.5} = -0.55$ for a neutron; $g = \frac{+5.79}{3.5} = 1.65$ for a proton if quenching effects are ignored). The fact that you need at least two particles before the $f_{5/2}$ configuration sets in, implies that the $f_{5/2}$ admixture renormalizes the magnetic moment operator by adding to it a two-(or more) body part, without changing the one-body part:

$$\mu = g\,J \rightarrow g\,J + \sum_{i<j} \delta\,\mu(i,j).$$

That is, if we want to pretend that our wave function is $f_{7/2}{}^n$, suppressing the $f_{7/2}{}^{n-1}\ f_{5/2}$ admixture we must use a combination of one body and two body operator, as above.

We can construct such a two body part explicitly as the product of a two-body scalar, something like a two-body potential and a one-body vectors

$$\delta\mu(12) = U(12)\ [J_z(1) + J_z(2)]$$

where J_z is the usual total angular momentum operator. Then

$$\langle j^{2I}\ \delta\mu(12)\ j^{2I}\rangle = I\ \langle j^{2I}\ U(12)\ j^{2I}\rangle.$$

If $f_{5/2}$ mixing is the only mechanism then the above is proportional to $\langle f_{7/2}{}^2\ I \mid V(12) \mid (f_{7/2}\ f_{5/2})\ I\rangle$, see Figure 1.

We need to know the two-body matrix elements only for I = 2, 4 and 6. We can either parameterize these three numbers or we can calculate them explicitly from the $f_{5/2}$ admixture. There are of course more fundamental sources of a two-body moment such as mesonic exchange corrections, etc.; unfortunately they get lumped together by the $f_{5/2}$ mixing and become obscured by it.

It has been pointed out to me by J. MacDonald and N. Koller[3] that the g factors for neutrons decrease in magnitude systematically as one goes up the calcium isotopes, and that the g factor for protons increases as we remove more protons

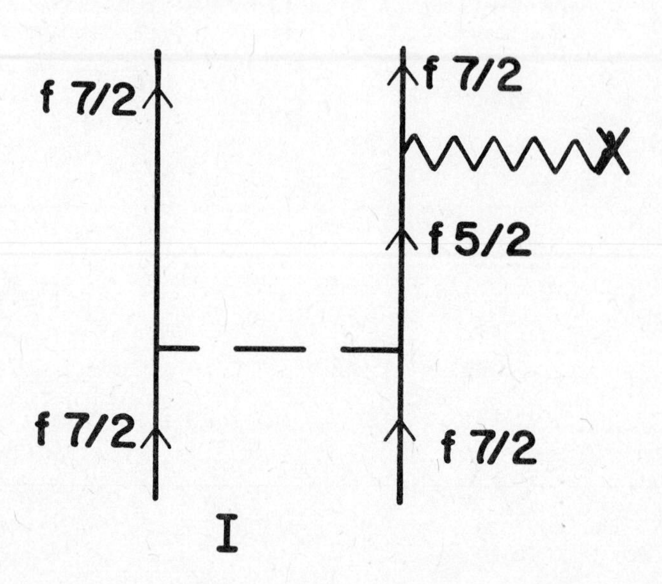

Fig. 1. Building block for a two-body magnetic moment operator in the $f_{7/2}$ shell.

from a Ni^{56} core. They present a nice table, see Figure 2, which has the following results:

a) neutrons $g(Ca^{41}) = -0.455$, $g(Ca^{42} J{=}6^+) = -0.42 \pm 0.025$ as measured by the Stony Brook group, $g(Ca^{43}) = -0.376$.
b) protons $g(Co^{55}) = 1.22 \pm 0.08$, $g(Fe^{54} 6+) = 1.35 \pm 0.07$ as measured by MacDonald, Hensler, Tape and Koller[3] at Rutgers-Bell, $g(Mn^{53}) = 1.43$, $g(Vn^{51}) = 1.46$.

In the $f_{7/2}$ model space this change with g is a pure two-(or more) body effect. Using a Kuo-Brown interaction[6] I have calculated the change of g for the protons due to $f_{5/2}$ admixture in perturbation theory. I used Mn^{53} rather than Co^{55} as the standard and I obtain

$$f(Co^{55}) = 1.27 \quad g(Fe^{54} 6+) = 1.35 \quad g(Mn^{53}) = 1.43 \quad g(V^{51}) = 1.60$$

which is quite reasonable.

If $f_{5/2}$ mixing is the only mechanism that is important then a simple relationship between the change in g for the protons and for the neutrons should hold:

$$\frac{(\delta g)\text{ protons}}{(\delta g)\text{ neutrons}} = \frac{-(g_\ell - g_s)\text{ protons}}{(g_\ell - g_s)\text{ neutrons}} \approx \frac{5.79-1}{3.83} = 1.25.$$

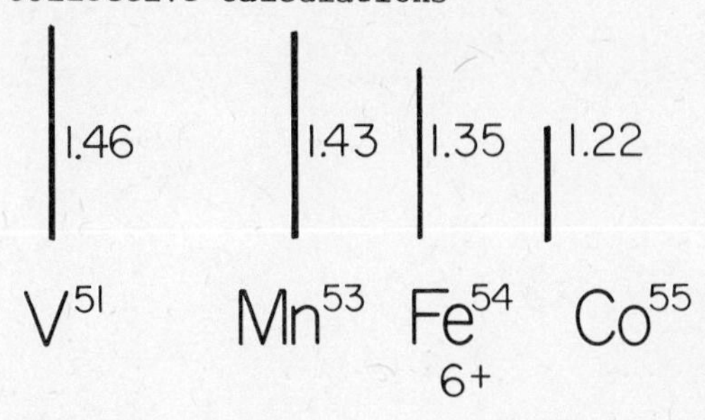

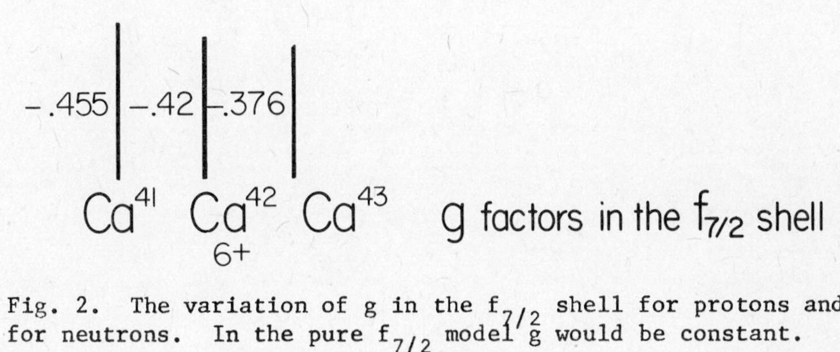

Fig. 2. The variation of g in the $f_{7/2}$ shell for protons and for neutrons. In the pure $f_{7/2}$ model g would be constant.

This holds fairly well. From the point of view of particles coupled to vibrations we note that the 1^+ T = 1 vibration in $Ni^{56} = \sum_i \sigma_\alpha(i)\, t_z(i) \mid Ni^{56} >$ is also a unique shell model state $f_{7/2}^{-1} f_{5/2}$. As we remove protons in going from Co^{55} to V^{51}, we are destroying the core which is necessary for the building of the vibration. It is then not surprising that g gets bigger. (Remember that in Ca^{40} the operator $\sigma_\alpha t_z \mid Ca^{40} > = 0$ in the shell model).

A second example, this is purely experimental, is the variation in charge radius of the calcium isotopes. In zero order, all isotopes Ca^{40} to Ca^{48} should have the same radius. In first order, Ca^{41} can polarize the core i.e. a small admixture of $f_{7/2} \times 0^+$ *MONOPOLE VIBRATION* would mix in. The radius of Ca^{41} would differ from that of Ca^{40}. Let $r^2(Ca^{41}) - r^2(Ca^{40}) = C$ (where r^2 is the square of the root mean square radius). Both first order perturbation theory and the intuitive ideas would lead to

$$r^2(Ca^{40+n}) = n\,C,$$

i.e. the charge in the square of the radius would be linear in n. But evidently this is not so. According to Frosch, *et al.*[4] we have (see Figure 3)

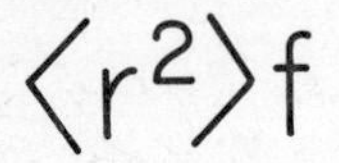

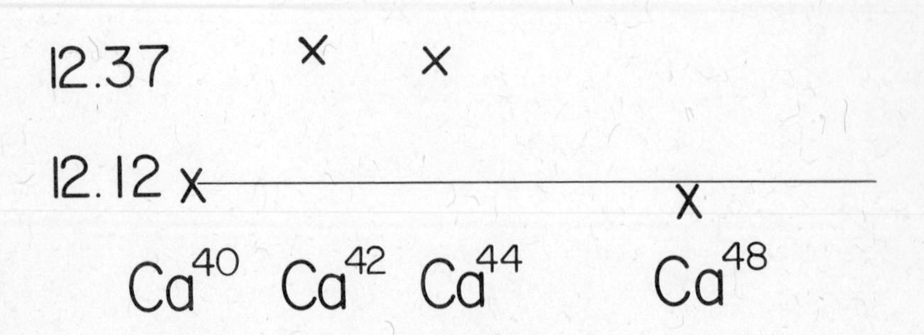

Fig. 3. The variation of the square charge radius through the calcium isotopes.

	$<r^2>^{\frac{1}{2}}$	$<r^2>$	$\delta r^2(40+n)$
Ca^{40}	3.487	12.12	0
Ca^{42}	3.517	12.37	0.21
Ca^{44}	3.515	12.35	0.20
Ca^{48}	3.477	12.08	-0.07

This nonlinear behavior in r^2 means that somehow the effective radius operator has a two-body part

$$\delta r^2 \text{ eff} = \sum_i O(i) + \sum_{i<j} V(i,j)$$

where the symbol V for the two-body radius operator has been written to suggest the similarity with the two-body potential, since both are scalars.

The problem of evaluating the two-body contribution in a given Calcium isotope $<j^{nJ} \sum_{i<j} V(ij)\, j^{n-J}>$ is identical to the problem of the binding energies of the calcium isotopes, and has been worked out by the Israeli group[5] as follows:

$$r^2\ (40 + n) = n\,C + n\frac{(n-1)}{2}\,\alpha + [\tfrac{n}{2}]\,\beta$$

$[\frac{n}{2}]$ = n/2 for even n; (n - 1)/2 for odd n.

The Stanford data[4] can be fitted with $\alpha = -0.041\ f^2$ and $2\,C + \beta = 0.233\ f^2$. To disentangle β from C requires a measurement of an odd Calcium isotope.

The third example of this will be the quadrupole operator. If I assume $f_{7/2}{}^2$ as the configuration of the 0_1^+ ground state of Ca^{42} and likewise the 2_1^+ state at 1.4 MeV

$$B(E2)\ 0_1 \rightarrow 2_1 = 113.5\ e_n^2\ (e^2\ f^4)$$

$$Q_{2_1^+} = 7.53\ e_n\ (e\ f^2)_1.$$

Allowing the two neutrons to be anywhere in the 2p-1f shell, Kuo and Brown[6] obtain the following wave functions

$$0_1^+ = 0.93\ f_{7/2}^{\ 2} + 0.20\ p_{3/2}^{\ 2} + 0.20\ f_{5/2}^{\ 2} + 0.11\ p_{1/2}^{\ 2} - 0.22\ g_{9/2}^{\ 2}$$

$$2_1^+ = 0.94\ f_{7/2}^{\ 2} + 0.28\ f_{7/2}\,p_{3/2} + 0.09\ p_{3/2}^{\ 2} + 0.05\ p_{3/2}\,f_{5/2} + 0.07\ p_{3/2}\,p_{1/2} + 0.96\ f_{5/2}^{\ 2} + 0.07\ f_{5/2}\,p_{1/2} - 0.16\ g_{9/2}^{\ 2}.$$

With these wave functions I calculate

$$B(E2)\ 0_1 \rightarrow 2_1 = 160.9\ e_n^2\ e^2 f^4$$

$$Q_{2_1^+} = 1.13\ e_n\ (e\ f^2).$$

The point I wish to make is that as a result of configuration mixing the B(E2) gets enhanced, but the quadrupole moment gets reduced drastically.

Thus we *cannot* simulate the effect of this configuration mixing by merely changing the effective charge: the polarization of the core by two nucleons is *not* the sum of the terms where each nucleon independently polarizes the core. There is a correction term which can be represented as a simple Feynman diagram as shown in Figure 1.

It is quite clear from this example, and this was first pointed out to me by Phil Goode in another context, that the $f_{7/2}^{\ 2}$ space is too small for properly doing calculations of 2_1^+ quadrupole moments. It might be possible to stay in $f_{7/2}^{\ n}$ with an effective two-body operator, but it is doubtful if this will be simpler than explicitly introducing configuration mixing.

Incidentally, the 2_1^+ state should have a large deformed admixture which has a negative quadrupole moment. It may well be that the measured value for Ca^{42} will be opposite of the shell model prediction.

Bob Lawson has pointed out to me that I should have been talking about Ti^{50} instead of Ca^{42}. In first order perturbation theory the two problems are identical. The nucleus Ti^{50} consists of two $f_{7/2}$ protons and a closed $f_{7/2}^{\ 8}$ shell of neutrons. According to Lawson, measurements have been done on this nucleus which do indeed show a *much* smaller quadrupole moment for the 2_1^+ state than one would obtain if one used an effective charge deduced from the B(E2) $2_1^+ \rightarrow 0_1^+$.

III. VIBRATIONS IN CLOSED SHELL NUCLEI

I define a vibration as an electromagnetic multipole operator acting upon the ground state of a nucleus. I immediately modify this definition so that the vibrations will carry a definite isospin as well as angular momentum. So we have isoscalar and isovector vibrations. As an example consider the electric monopole isoscalar:

Let $|o\rangle$ be O^{16} ground state in the shell model.

$$\text{Then } \psi^{J=0T=0} = \sqrt{3/18}\,[2\,s_{1/2}\;1\,s_{1/2}^{-1}]^{J=0T=0}$$

$$+ \sqrt{10/18}\,[2\,p_{3/2}\;1\,p_{3/2}^{-1}]^{J=0T=0}$$

$$+ \sqrt{5/18}\,[2\,p_{1/2}\;1\,p_{1/2}^{-1}]^{J=0T=0}.$$

With a harmonic oscillator single particle hamiltonian, the unperturbed position of this state is $2\hbar\omega$. From the measured rms. radius of O^{16} $\langle r^2\rangle^{1/2} = 2.67f^4$ we deduce $2\hbar\omega = 83 \times 9/2\langle r^2\rangle = 26.2$ MeV (for Ca^{40} $\langle r^2\rangle^{1/2} = 3.43$ $2\hbar\omega = 83 \times \frac{6}{\langle r^2\rangle} = 20.9$ MeV). The mean energy of this state due to the residual interaction is $\hbar\omega$ + V part-core + V hole-Core + V part-hole where the sum of the first 3 terms would be $2h\omega$ if by some miracle self consistency had been attained.

However, the trend of (p, 2p) experiments[9] and Hartree-Fock calculations[10], indicates that the 1s level is bound much deeper than is given by the harmonic oscillator. This is shown in the following two tables.

TABLE I

Comparison of $2\hbar\omega$ versus $\hbar\omega$ + V part-core + V hole-core

State	Sussex Matrix Elements[a)] $2\hbar\omega = 28.8$ MeV	Soper Force[b)] $2\hbar\omega = 28.6$ MeV
$2\,S1/2 - 1\,S1/2^{-1}$	43.8	40.7
$2.P3/2 - 1\,P3/2^{-1}$	32.3	29.6
$2\,P1/2 - 1\,P1/2^{-1}$	29.8	26.6

a) As calculated by H. Mavromatis[7], who emphasized the importance of this.

b) As calculated by R. Mohan[8]. $V_{Soper} = -560\,(0.865 + 0.135\,\sigma_1 \cdot \sigma_2)\,\delta(\vec{\gamma}_1 - \vec{\gamma}_2)$ MeV.

TABLE II

PROTON SEPARATION ENERGIES

	State	Experiment[a)]	Harmonic Oscillator $\hbar\omega = 14$	Negele's[b)] Hartree-Fock
O^{16}	1 S	-40 ± 8	-28	-32.8
	1 P	-16.8	-14	-14.2
			$\hbar\omega = 11$	
Ca^{40}	1 S	-50 ± 11	-32.	-43.2
	1 P	-34 ± 6	-21.	-26.9
	1 d	-12.6	-10.	-11.8
	2 S	-11.61	-10.	- 8.6
	1 f	- 1.1	-1	- 0.2

a) The deep levels were obtained in the (p,2p) experiment of A.N. Janes et al.[9]

b) Starting with a realistic Reid soft core potential, a density dependent G matrix was derived and used in a Hartree-Fock calculation. Density independent forces tend to get the 1s much much deeper. See Ref. 10.

The particle-hole interaction for a monopole vibration is extremely attractive. For example, it is about -20 MeV with the Kallio-Kolltveit interaction[11] (with $\hbar\omega = 14.8$), it is -16.8 MeV with the Sussex matrix elements. Mavromatis[7] shows that if we choose the particle-hole splitting at $2\hbar\omega$ to be 28.8 MeV the monopole state comes down to 12 MeV above ground, if one calculates Vp core + V_h core it comes at 19.7 MeV.

Next we come to calculations in the random phase approximation. We calculate mean energy as before, but we use more sophisticated ground state wave functions, which include ground state correlations. In first order perturbation theory, assuming self consistency, one will admix two particle-two hole configurations into the ground state via the residual interaction

$$\psi_{6.5.} = |o\rangle + \sum_{2P-2H} \frac{\langle o\ V\ 2P\text{-}2H\rangle}{E_o - E_{2P-2H}} \psi_{2P-2H}.$$

The full R.P.A. correlations are more elaborate. They have 4p-4h, 6p-6h etc., but in the limit of small V they reduce to the above. For a long time there was an erronious expression for the ground state correlations - a familiar exponetial form which did not reduce to perturbation theory - it predicted twice the amount of correlation - but we now know this formula is wrong.

When the ground state correlations are included, the monopole state comes embarrassingly low - things are worse in calcium than in oxygen. In fact with the single particle splittings at $2\hbar\omega$ and using some standard realistic forces these vibrations come below the ground state in R.P.A. calculations. This was first noted by Blomqvist[12], but we shall quote results of a preprint by P. Goode, B. J. West and S. Siegel[13] who find in Ca^{40}, (with the Kuo-Brown force and using $2\hbar\omega$ = 22 MeV for the single particle energies), that the monopole state should be 4.14 MeV above ground with no ground state correlations (T.D.A.) but is imaginary in the R.P.A. This is shown in Table III. Only by choosing very high single particle excitation energies for *all* single particle excitations, can they get a non-imaginary value for

TABLE III

Excitation Energy of the T = 0 Monopole State in Ca^{40}[a]

Remark	T.D.A. Energy (MeV)	R.P.A. Energy (MeV)
K. K. Force, no ρ $\hbar\omega + V_{pc} + V_{Hc}$ = 21 MeV in all states	1.17	Imaginary
Kuo - Brown Force $\hbar\omega + V_{pc} + V_{Hc}$ = 21 MeV in all states	4.14	Imaginary
Kuo - Brown $\varepsilon_{1f-1p} = 42$ $\varepsilon_{35-25} = 21$	8.02	Imaginary
Kuo - Brown $\varepsilon_{1f-1p} = 42$ $\varepsilon_{35-25} = 42$	24.4	20.9

a) P. Goode, B. J. West, Stan Siegel, U. of Rochester report UR-875-318

the breathing mode energy in the R.P.A.

Following a suggestion of Kuo and Blomqvist, I have found that the R.P.A. is not a good approximation for 0^+ vibrations, especially in light nuclei. For example in O^{16}, with the Kallio-Kolltveit force[1] I found (refer to Figure 4):

1. In the T.D.A. the particle hole force for the monopole state is -20 MeV.

2. Including the R.P.A. type ground state correlation theory, this means admixing 3P-3H, gave an additional -2.5 MeV.

3. But, the inclusion of a non-R.P.A. correlation, this corresponds to the admixture of a 2p-2h state, gave a contribution in the opposite direction + 4.2 MeV. This is called a shielding correction by Kuo and Blomqvist[14]. Gerry Brown and Georges Ripka[18] have noted that some of these non-R.P.A. corrections are included automatically if one uses a density dependent G-matrix, but I don't have time to go into this.

It should be noted that there exists another approach, completely phenemological, for calculating the breathing mode energies. This is the hydrodynamical model. Walecka[15] and others showed that in this the breathing mode energy is proportional to the square root of the nuclear compressibility. Let us follow the recent work of Pandharipande[16]. He writes the relationship

$$(\hbar\omega)_{\substack{\text{breathing}\\ \text{mode}}} = A^{-1/3}\ \frac{\lambda CN}{2\pi r_o}\ (k\ M_n C^2)^{1/2}$$

where λ_{CN} is the nuclear compton wave length, r_o the unit volume radius in nuclear matter and k, the compressibility is defined:

$$k = r_o^{1/2}\ \frac{1}{A}\left[\frac{\partial^2 E(r_e)}{\partial r^2}\right]_{\substack{\text{at}\\ \text{equilibrium}}}$$

$$r_e = A^{-1/3}\ (5/3)\ \langle r^2\rangle^{1/2}.$$

This puts the problem in better perspective. Apparently when we nuclear shell model people are doing a casual particle-hole calculation, we are (implicitly) doing no less than calculating the second derivative of the binding energy with respect to the radius. Remembering that it is a hard enough problem to get the binding energy and the density of a nucleus to come out right with realistic forces, the second derivate should be that much more difficult.

The compressibility is not known experimentally. All theoretical estimates are such that the breathing mode should come at a high energy. Pandharipande[16] shows that compressibilities in finite nuclei should be smaller than in nuclear

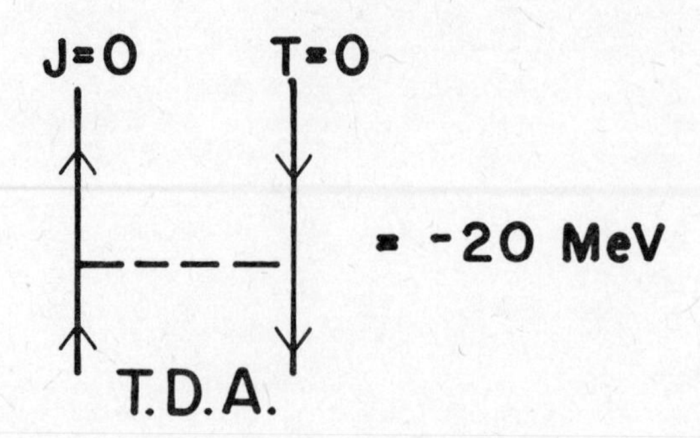

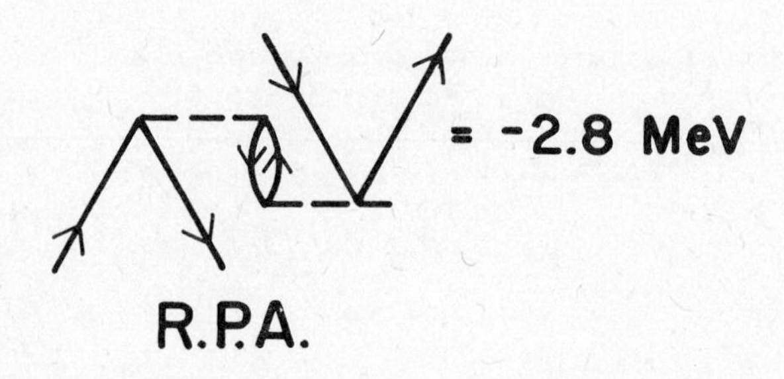

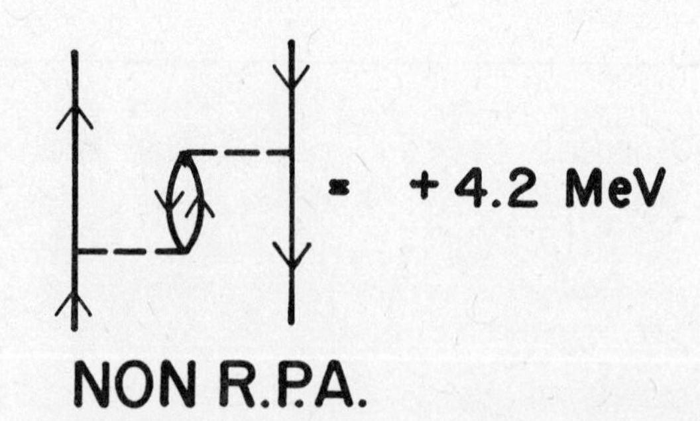

GRAPH "SCREENING"

Fig. 4. Contributions to the energy of the monopole isoscalar (or breathing mode) state, for O^{16} using the Kallio-Kolltveit interaction

matter; he gets for Ca^{40}, k = 123 E breathing mode = 18.3 MeV. A similar calculation of Breuckner *et al.*[17] gets it even higher E_{BM} = 29.5 MeV in Ca^{40} and 30.5 MeV in O^{16}.

IV. PARTICLES COUPLED TO VIBRATIONS

Effective Charge and Core Renormalization of the Two-Body Force.

To get a finite quadrupole moment for say the J = $7/2^-$ ground state Ca^{41} we must admix into the basic configuration $f_{7/2}$ | closed shell > some component in which the core is excited - the simplest such core excitation consists of a one particle - one hole excitation. Only if the P-H excitation is a proton excitation with spin 2 and positive party will we get a finite moment. Thus the P-H excitation must be a $2\hbar\omega$ excitation e.g. $1p_{1/2}^{-1}\ 1f_{5/2}$, $1d_{5/2}^{-1}\ 1g_{9/2}$ etc. The effective charge in first order, e°_N is the ratio of the quadrupole moment calculated in first order to the quadrupole moment of an $f_{7/2}$ proton. We can do the same thing for Sc^{41} and get e°_p. It is convenient to define the isoscalar and isovector charges e(T = 0) = $e_N + e_p$ e(T = 1) = $e_p - e_N$. Experimentally e_N = 1.3 and e_p = 0.7 for Ca^{41} and Sc^{41} from the $2p_{3/2} \rightarrow 1f_{7/2}$ transitions. The presence of deformed states obscures this analysis, but Gerace and Green[19], who include these, still need e_N = 1 and e_p = 0.5 for the single particle transition.

We can also get the effective charge by constructing a quadrupole vibration in Ca^{40} $\psi^{2+}_{quad} = \sum_{protons} r^2 y^2$ | ground state > and coupling an $f_{7/2}$ particle to it $f_{7/2} \times \psi^{2+}_{quad}$. If we assume that the state ψ^{2+}_{quad} is at its unperturbed energy ($2\hbar\omega$ in the oscillator model) and we admix this state perturbatively we get exactly the same as perturbation theory. The effective charge problem is similar to the core renormalization of the two-body force. In fact with a quadrupole-quadrupole force, the ratio of the core renormalized part of the two body force to the bare matrix element G3p-1h/GBARε is precisely the *isoscalar effective charge*. This is shown schematically in Figure 5.

With a more general force the particle-hole transferred between two nucleons need not have only quantum numbers J" ± equal two but in fact any J" allowed consistent with angular momentum conservation.

Using ε particles - ε hole = $2\hbar\omega \approx$ 21 MeV Federman and I[20] found that the effective charge in first order was too small, e_N = 0.6 and e_p = 0.2. Then Siegel and I[21] did some R.P.A. calculations. This consisted of summing all diagrams in which the bubble is iterated backwards and forwards, as is shown in Figure 6. The R.P.A. has a very simple struc-

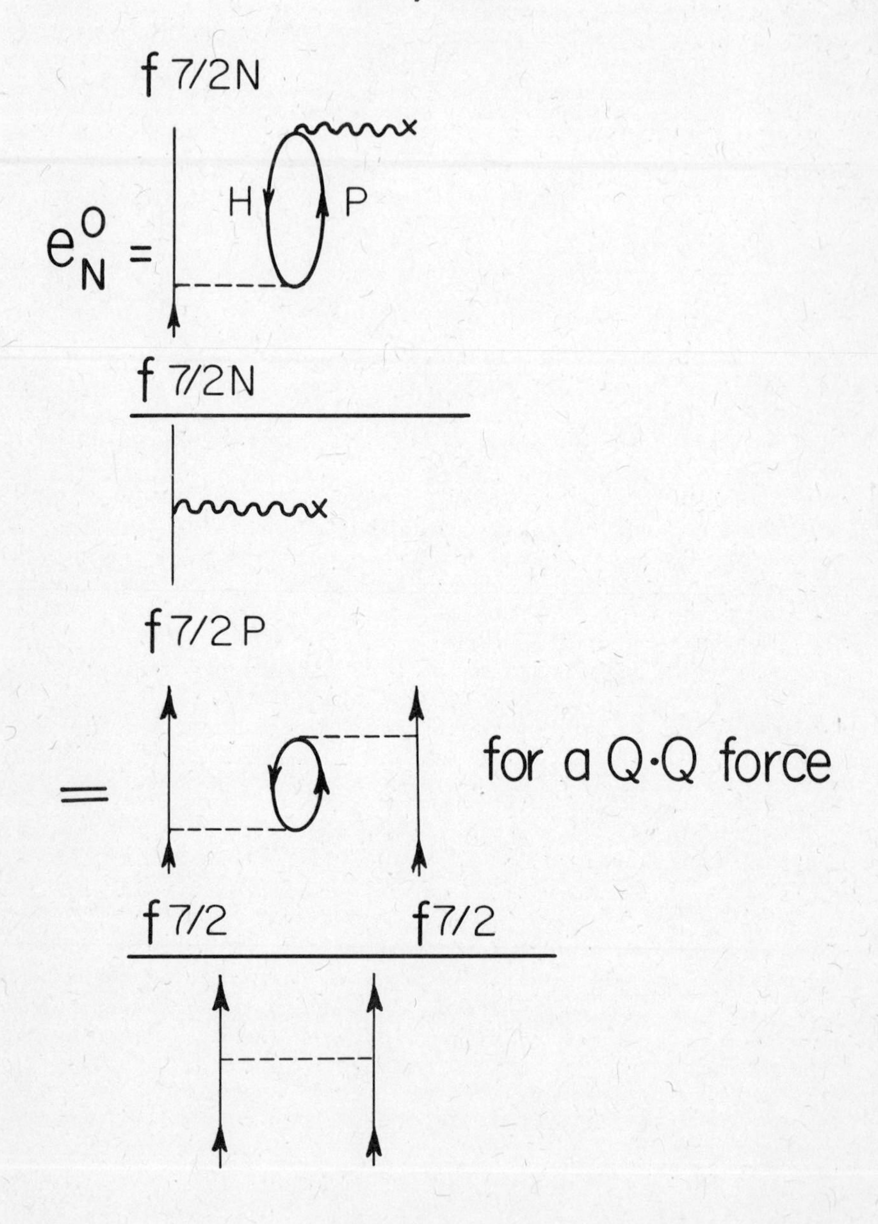

Fig. 5. With a quadrupole-quadrupole force the core renormalization of the 2-body force is simply related to the isoscalar effective charge.

ture. You get uncoupled equations for the isoscalar and isovector effective charges. If the force used is the quadrupole quadrupole force then there is a very simple relationship between the R.P.A. charge and the 1st order charge.

$$e(T=0)_{R.P.A.} = \frac{e^{\circ}(T=0)}{[1 - e^{\circ}(T=0)]} .$$

R.P.A. SERIES

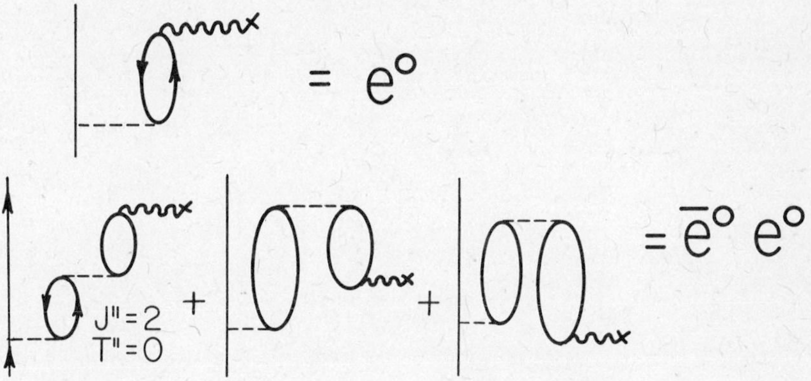

$= e^{\circ}$

$J'' = 2$
$T'' = 0$

$+ \quad + \quad = \bar{e}^{\circ} e^{\circ}$

$$e = e^{\circ} + \bar{e}^{\circ} e^{\circ} + (\bar{e}^{\circ})^2 e^{\circ} + \ldots$$

$$= \frac{e^{\circ}}{1 - \bar{e}^{\circ}}$$

For a Q·Q force $\bar{e}^{\circ} = e^{\circ}$

For "Realistic" force $\bar{e}_{\circ} > e^{\circ}$

Fig. 6. First order and the R.P.A. series for the effective charge.

In other words the R.P.A. is a geometric sum in which the first order isoscalar effective charge is the expansion parameter. But note that in first order Federman and I[20] got $e^\circ(T=0) = 0.6 + 0.2 = 0.8$. If we put this in the R.P.A. series we get $e = \frac{0.8}{1-0.8} = 4$. We are dangerously close to an instability.

Essentially the same results occur when one uses the realistic force rather than a quadrupole force. The structure now becomes

$$e(T) = \frac{e^\circ(T)}{1-\bar{e}^\circ(T)}$$

where e° is the effective charge due to the valence nucleon polarizing the core and $\bar{e}^\circ$ is the effective charge due to a particle - hole pair polarizing the core. In general we found that $\bar{e}^\circ$ was even greater than e°. The same enhancement occurs for the core renormalization of the two-body force. There are several mechanisms all of which go in a direction of removing this instability:

1). More careful consideration of the energy denominators, as was discussed in the case of the monopole vibration, it applies to the quadrupole case as well. For example, Goode, West and Siegel[13] get the 2^+ quadrupole state in Ca^{40} at 8.27 MeV with the Kallio-Kolltvelt force. Is the mean energy of this state so low? Probably. The instability that Siegel and I[21] got was due to the isoscalar quadrupole state in Ca^{40} coming down very low.

2). State dependence of the effective interaction. The average kinetic energy of a nucleon which has been excited through 2 major shells is larger than that of a nucleon in the core. If two nucleons have a higher relative velocity, then they should feel the hard core more and so the average interaction should be less attractive. This effect makes $\bar{e}(PH)$ (the renormalization by an excited particle-hole pair) smaller.

3). Perhaps the R.P.A. is not so good? Siegel[21] studied this in O^{17}, F^{17} in second order perturbation theory and found that if one includes only the R.P.A. graphs up to second order (for a $d_{5/2}$ nucleon) one finds that the isoscalar effective charge is 0.93. If one includes all second order diagrams the isoscalar effective charge is 0.72. In first order the isoscalar charge is only 0.55. So we see, complete second order is larger than first order but considerably smaller than the R.P.A.

4). Density dependent forces: H. Bethe has advocated the use of a force which is weaker inside the nucleus where the density is large than it is outside the nucleus where the density is low. A poor man's version of such a force is the A. M. Green modified Kallio-Kolltveit potential[22]

$$V\ (\text{singlet}) = 0.99[1 - \beta\ 0.06\ \rho(R)]\ V_{kk}\ (\text{singlet})$$

$$V\ (\text{triplet}) = 1.07\ [1 - \beta\ 2.65\ \rho(R)]\ V_{kk}\ (\text{triplet})$$

$$\beta = 1, \qquad \rho(R) = \rho_o\ \{1 + \exp\ [(R - C)/a]\}^{-1}$$

This potential is much weaker inside the nucleus than Kallio-Kolltveit and is equal to it outside.

Obviously since it is weaker than Kallio-Kolltveit it will lead to smaller effective charges. This is shown explicitly in the work of Goode, West and Siegel[13]. Gerry Brown and independently Georges Ripka[18] caution that it is wrong simply to use a density dependent force in a calculation. There are terms in $\rho \frac{\partial v}{\partial \rho}$ which must also be included. The inclusion of these terms make the potential even weaker than before and consequently a smaller effective charge. Roughly speaking change β in the density dependent force, above, from 1 to 2 and this will take the $\rho\frac{\partial V}{\partial \rho}$ terms into account. (If I remember the prescription of Ripka correctly, the particle-hole force is roughly $V + (3/4)\ \rho\ \partial V/\partial\rho$).

5). Full Hartree-Fock calculations for a closed shell plus one nucleon yield deformed solutions, from which one can obtain the quadrupole moment and consequently the effective charge. A consistent feature of these calculations which employ semi-realistic soft core potentials is that they yield very *small* effective charges (private communication of Cusson and M. Harvey). Why is this? I am not sure. In a perturbation theory approach, we are also presumably calculating the change in the Hartree-Fock field due to the addition of one nucleon and so it seems that we are doing the same thing. But Cusson gets $e^{\circ}_{N} = 0.15$ whereas in first order perturbation theory I got 0.41. Maybe it is because in Cusson's calculation the $1s_{1/2}$ comes very low i.e. - 80 or - 90 MeV. Maybe the implicit energy denominators are then very large. M. Harvey has suggested that a further study of the relationship between perturbation theory and the Hartree-Fock would be of great interest.

V. NEGATIVE PARITY VIBRATIONS AND PARTICLES COUPLED TO THEM

The lowest 3^- and 5^- in Ca^{40} at 3.7 MeV and 4.5 MeV respectively are called collective because their excitation energies are lower than the $f_{7/2}$ $d_{3/2}$ single particle splitting of 7.2 MeV, and because the respective E3 and E5 decays to ground are very strong. The two other members of the $(f_{7/2}\ d_{3/2})^{-1}$ multiplet $J = 2^-$ and 4^- are at higher energies and probably will not have strong decay properties to ground.

The 3^- state consists mostly of $1\hbar\omega$ particle-hole excitations and cannot therefore be identified with $\Sigma r^3 y^3 |o\rangle$, the latter consisting of a great deal of $3\hbar\omega$ excitations.

In fact Gillet and co-workers, limiting themselves to 1 $\hbar\omega$ excitations were able to find a particle-hole force which would fit the energies and decay strengths of these states, provided they introduced ground state correlations (R.P.A.), if they treated Ca^{40} as a closed shell (T.D.A.) the decay rates were a factor of five or so too slow. The resulting wave functions, however, appeared to have much more configuration mixing than was seen in one nuclear transfer reaction. It was subsequently shown that there is a small admixture of $3\hbar\omega$ excitations in the 3^- state, can cause an enhancement of about *two* in the E3 decay, thus allowing for the reduction of configuration mixing in the 1 $\hbar\omega$ part of the particle-hole excitations.

We briefly mention realistic forces. We note that the matrix element $\langle[f_{7/2}\, d_{3/2}^{-1}]^{3-T=0}\, V[f_{7/2}\, d_{3/2}^{-1}]^{3-T=0}\rangle \approx 0$ with the Kuo-Brown force[6], so the lowering of the 3^- state has to come from configuration mixing. The calculation is a bit tricky because whenever a state is collective it becomes supersensitive to the parameters. For example Dieperink et al.[23] get the state to be unstable if they use the Tabakin force with both 1 $\hbar\omega$ and 3 $\hbar\omega$ excitations allowed (although with only 1 $\hbar\omega$ excitations it comes at 2.4 MeV). Also Kuo and Blomqvist[14] showed that non R.P.A. effects such as shielding are very important and go in the opposite direction of the R.P.A. i.e. they raise the energy of the 3^- state. So everything is a delicate balance, some realistic forces will get it too low, some too high, the simple R.P.A. cannot be trusted. M. Kirson[24] is making an extensive study of this problem, but I do not yet have his results. As one looks at the results of many different calculations one cannot help but notice that this lowest 3^- state seems, to a very good approximation to obey an energy weighted sum rule all by itself.

$$[E\ B(E3)]^{3^-}_{\text{for one given force}} = [E\ B(E3)]^{3^-}_{\text{for another force}}.$$

So any force that gets the 3^- state too low gets the E3 decay rate too high, and vice versa.

Whereas in heavy nuclei such as Bi^{209} one has almost perfect examples of a particle coupled to one 3^- vibration, the situation around Ca^{40} is somewhat more complicated, in a certain sense more interesting.

Take K^{39} as an example, the shell model ground state consists of a $d_{3/2}$ proton hole coupled to a Ca^{40} core. In the weak coupling model one assumes that the low lying opposite parity states consist of a $d_{3/2}$ hole coupled to the lowest 3^- vibration of Ca^{40}. This leads to a multiplet of states $\vec{3}/2 + \vec{3}^- = 3/2^-, 5/2^-, 7/2^-, 9/2^-$ each occurring once.

Furthermore in the weak coupling model the E3 rates in K^{39} are simply related to those in Ca^{40} by the sum rule

$$B(E\ 3: K^{39})\ (\tfrac{3}{2}^{+} \rightarrow J)$$
$$= (2\ J + 1)\ \frac{B(E\ 3: Ca^{40})\ (0^{+} \rightarrow 3^{-})}{(2 \times \frac{3}{2} + 1)\ (2 \times 3 + 1)}\ .$$

Note that the following sum rule holds in this weak coupling model

$$\sum_J B(E3: 3/2 \rightarrow J) = B(E3: 0 \rightarrow 3).$$

As seen in Figure 7, the negative (opposite) parity spectrum in K^{39} is a bit more complicated than the weak coupling model prediction. There are two $7/2^-$ states, one at 2.82 MeV and the other at 3.94 MeV. But there is still a strong suggestion of particle x vibration. We see the $J = 3/2^-$, $5/2^-$ and $9/2^-$ members of the multiplet at 3.02, 4.15, and 3.60 MeV respectively.

In Figure 8 is the result of a calculation by Goode and me[25] for the opposite parity states of both K^{39} and Ca^{41}. In the middle of the figure we have Ca^{40} as a reference. First one needs the vibrations in Ca^{40}. Note that experimentally the 3^- and 5^- states are at 3.72 and 4.5 MeV respectively. The 4^- member of the $(f_{7/2}\ d_{3/2})^{-1}$ multiplet is at 5.4 MeV which is not really all that much higher. The 2^- member of this multiplet has never been seen (there is a 2^- state at about 6 MeV but this is mostly a 3p-3h state, it is a member of the rotational band of which the 1^- state is the lowest member. The 3^- state at 6.2 MeV is also a member of this rotational band and according to Goode only 6% of the deformed 3^- state is admixed into the 3.72 MeV (basically vibrational state).

Rather than choose a force which gave a good fit to the 3^- and 5^- vibrations, like Gillet had done, (and it's really not a bad idea to try such a fitting approach) we used the results of a calculation with the realistic Kuo-Brown[6] interaction as was performed by Gerace and Green[19]. The $f_{7/2}$ $d_{3/2}$ splitting was taken to be 7.2 MeV. Only 1 $\hbar\omega$ excitations were included.

In this calculation the 3^- and 5^- states come out at about 1 MeV too high. Not surprisingly then the opposite parity states come out about 1 MeV too high in K^{39} and Ca^{41} --this is nothing to be worried about.

Also listed in Figure 8 are the so-called T.D.A. energies, these are much higher, e.g. the $J = 3/2^+$ state in Ca^{41} comes at 4.5 MeV instead of at 2.02 MeV. To a large extent, it was a sense of frustration with the T.D.A. results which led us to consider the particle x vibration.

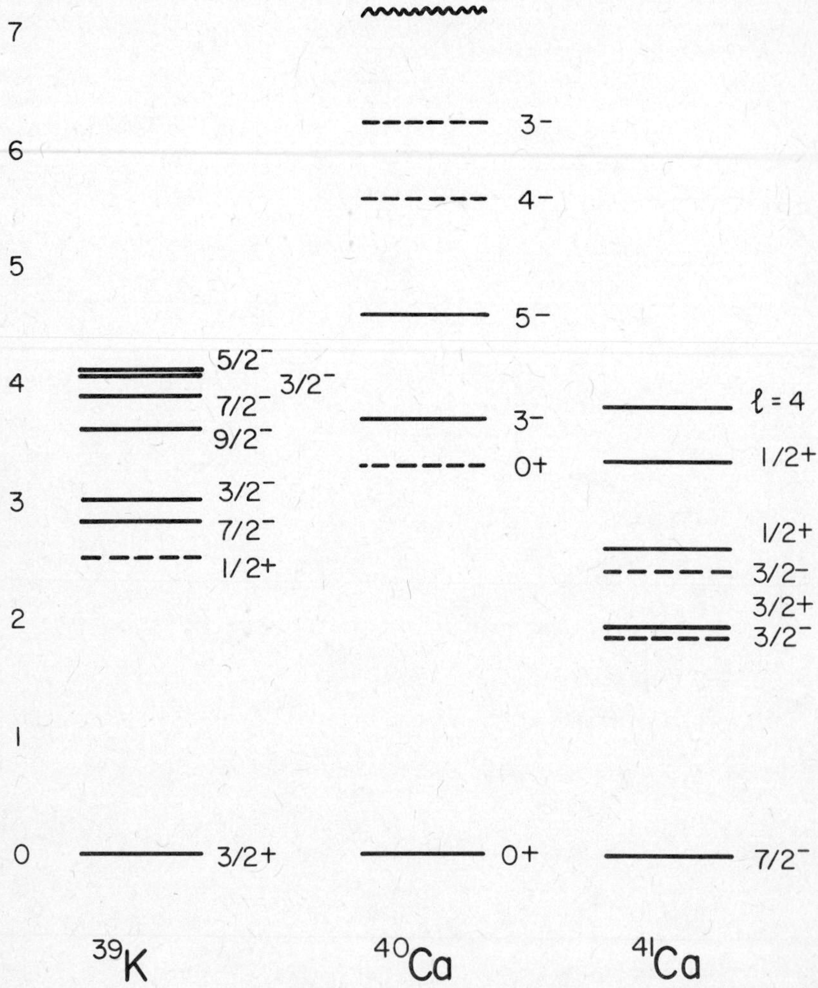

7
6
5
4
3
2
1
0

5/2⁻ 3/2⁻
7/2⁻
9/2⁻
3/2⁻
7/2⁻
1/2+
3/2+

3-
4-
5-
3-
0+
0+

ℓ = 4
1/2+
1/2+
3/2-
3/2+
3/2⁻
7/2⁻

^{39}K ^{40}Ca ^{41}Ca

Fig. 7. Selected states in K^{39}, Ca^{40} and Ca^{41}. The solid lines might be vibrational states.

By T.D.A. I mean just an ordinary shell model matrix diagonalization--1p-1h in Ca^{40}, 2p-1h in Ca^{41} and 2h-1p in K^{39}. Not only the absolute energies but also the relative energies, e.g. the $J = 1/2^+$ energy relative to the $J = 3/2^+$ energy are poorly given in the T.D.A. Note that in K^{39} the energy difference of the $1/2^+$ hole state and $3/2^+$ hole state is large → 2.5 MeV. In Ca^{41} it is much smaller only 0.7 MeV. We could not get this big change with the ordinary shell model diagonalization but in the R.P.A. it is readily available.

Our result was that the lowest $7/2^-$ state in K^{39} is mostly a $d_{3/2}$ hole coupled to a 5^- vibration and the second $7/2^-$ state mostly a $d_{3/2}$ hole coupled to a 3^- vibration. The other spins $3/2^-$, $5/2^-$ and $9/2^-$ involve mainly the 3^- vibra-

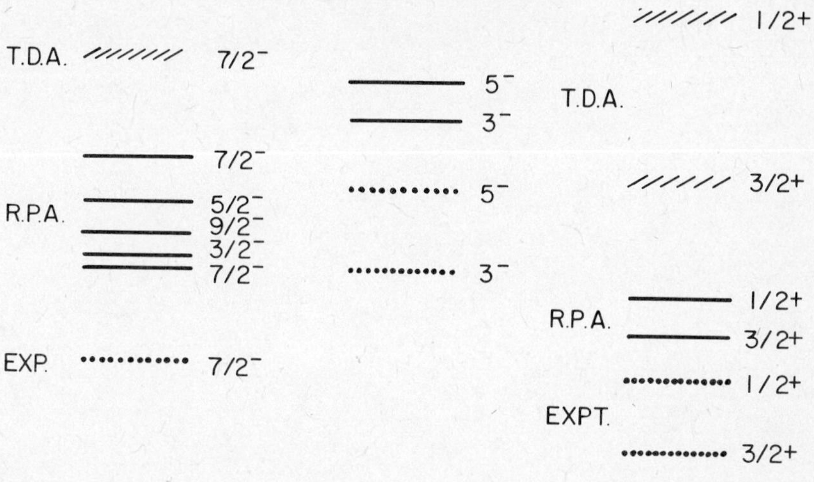

Fig. 8. A comparison of R.P.A. and T.D.A. (i.e., ordinary matrix diagonalization) for the hole (particle) X negative vibration of Ca^{40}, for K^{39} (Ca^{41}). The Kuo-Brown force (Ref. 6) was used and an $f_{7/2}-d_{3/2}$ splitting of 7.2 MeV was taken. The calculation was done by Phil Goode.

tion. Not worrying for the moment about lack of orthogonality of the basis states we get

$$|\psi_{7/2}{}^{-}\rangle = 0.44\,|\,(d_{3/2}{}^{-1}3^{-})^{7/2^{-}}\rangle - 0.88\,|\,(d_{3/2}{}^{-1}\,5^{-})^{7/2^{-}}\rangle$$

$$|\psi_{7/2_2}{}^{-}\rangle = 0.88\,|\,(d_{3/2}{}^{-1}\,3^{-})^{7/2^{-}}\rangle + 0.44\,|\,(d_{3/2}{}^{-1}\,5^{-})^{7/2^{-}}\rangle.$$

This result is consistent with the ($\alpha\alpha'$) reactions of R. J. Peterson[27] which excite the second $7/2^-$ state much more strongly than the first $7/2^-$ state. This is shown in Figure 9. Aside from exciting two $7/2^-$ states their results are consistent more or less with the weak coupling model.

Let me make some important remarks about a transition rate calculation. We assume only the two states $[d_{3/2}{}^{-1}\times 3^-]^{-J}$ and $[d_{3/2}{}^{-1}5^-]^{-J}$ are important. The result of solving the Schroedinger equation gives us, what Klein, calls amplitudes for each state $A^{L-J} = \langle\psi^{-J}[d_{3/2}{}^{-1}\psi^{L}_{vib}]^{J}\rangle$ where ψ^J is the state of K^{39} in question.

	E	J	B(E3)↑ ($\alpha\alpha'$)
^{40}Ca	3.90	3^-	24.4
^{39}K	2.82	$7/2^-$	1.0
	3.02	$3/2^-$	2.3
	3.60	$9/2^-$	6.3
	3.87	$7/2^-$	4.8
	4.11	$5/2^-$	5.1
			19.5

WEAK COUPLING MODEL:

$$B(E3)\,^{39}K^{J}\uparrow = (2J+1)\,\frac{B(E3)\,^{40}Ca^{3^-}\uparrow}{(2\times 3/2+1)\;(2\times 3+1)}$$

Fig. 9. Inelastic α-scattering in K^{39} from the $J = 3/2^+$ ground state to states of negative parity, which we describe here as a $d_{3/2}$ hole coupled to the 3^- or 5^- vibrations of Ca^{40}. The experiment was done by R. J. Peterson (Ref. 27).

It is tempting to identify A^{LJ} with the probability amplitude that a given state in K^{39} looks like $(d_{3/2}^{-1}\psi^L_{vib})$, but this is not correct because these states are not orthonormal. One sets up the normalization matrix

$$N^J(L,L') = \langle [d_{3/2}^{-1}\,\psi^L_{vib}]^J\,[d_{3/2}^{-1}\,\psi^{L'}_{vib}]^J\rangle$$

and diagonalizes it to obtain two orthonormal states $|I\rangle$ which are linear combinations of the original states

$$|I\rangle = \sum_L C^{I-J}_{\;L}\,[d_{3/2}^{-1}\,\psi_{vib}^{\;L}]^{-J}.$$

Now an EL decay is proportional to the square of the amplitude

$$\langle d_{3/2}{}^{-1}\, E^L_M\, \psi^J\rangle \text{ where } E^L_M = r^L\, y^L_M \,.$$

We write

$$\langle d_{3/2}{}^{-1}\, E^L_M\, \psi^J\rangle = \sum_{\pm} \langle d_{3/2}{}^{-1}\, E^L_M\, I\rangle \langle I\psi^J\rangle$$

$$= \sum_{\pm LL'} C^I_L\, C^I_{L'} \langle d_{3/2}{}^{-1}\, E^L_M\, [d_{3/2}{}^{-1}\, \psi^L_{vib}]^J\rangle \langle [d_{3/2}{}^{-1} A^{L'-J} \psi^{L'}]^J\, \psi^J\rangle.$$

Now the matrix element of E^L_m has its main contribution when $L = \mathcal{L}$. If we neglect the correction terms we get

$$\sum_{\mathcal{L}'} (\sum_I C^I_L\, C^I_{\mathcal{L}'})\, A^{\mathcal{L}'LJ} \langle d_{3/2}{}^{-1}\, E^L_M\, [d_{3/2}{}^{-1}\, \psi^L_{vib}]^J\rangle.$$

Note that even though we are considering a specific EL transition, say L = 3 the amplitude for this is a coherent sum over two $\mathcal{L}$'s corresponding to coupling of the hole to a 3^- vibration ($\mathcal{L}$ = 3) and 5^- vibration ($\mathcal{L}$ = 5).

It turns out, that in the $7/2_1 \rightarrow 3/2$ transition there is destructive interference between two amplitudes, whereas for the $7/2_2^- \rightarrow 3/2^+$ transition the interference is constructive. This is in agreement with experiment.

Another point of interest, in the (α,α') reaction from say $3/2^+_1 \rightarrow 7/2_1^-$ one should see both the E3 and E5 parts (they are incoherent for unpolarized beams). However, Peterson could not detect any E5 part to either the $7/2_1^-$ or $7/2_2^-$ states. This is worrysome and has not been resolved.

The calculations were also performed in Ca^{41} and Ca^{42}. In Ca^{42} in particular, Goode[26] obtained a much richer spectrum than was obtained in a shell model calculation.

ACKNOWLEDGEMENTS

I would like to acknowledge the help I got from Mike Kirson, Phil Goode, Jack MacDonald, Naomi Koller, Paul Ellis, Radbe Mohan and Stan Siegel.

REFERENCES

1. A. Arima and H. Horie, Prog. Theor. Phys. 11, 509 (1954).

2. S. Fallieros and R. A. Ferrel, Phys. Rev. 116, 660 (1959).

3. J. R. MacDonald, R. Hensler, J. W. Tape, and N. Benczer-Koller, private communication and Bull. Am. Phys. Soc. 15, 1666 (1970).

4. R. F. Frosch, R. Hofstadter, J. S. McCarthy, G. K. Noldehe, K. V. van Oostrum, B. C. Clark, H. Herman and R. G. Ravenhall, Phys. Rev. 174, 1380 (1968). See also R. P. Singhal, J. R. Morieta and H. S. Caplan, Phys. Rev. Letters 24, 73 (1969). They say that the radius of O^{17} is less than that of O^{16}, and the radius of O^{16} is less than that of O^{18}.

5. For example, I. Talmi, Rev. Mod. Phys. 34, 704 (1962)

$$\alpha = -(2(2+1)\vec{E}_2 - E_0)/(2j+1)$$
$$\beta = 2(j+1)/(2j+1)(\vec{E}_2 - E_0)$$

where $E_0 = \langle j^2 J = 0 | V | j^2 - J = 0 \rangle$

$$\vec{E}_2 = \sum_{j \neq 0} (2J+1) \langle j^2 J | V j^2 - J \rangle / \sum_{j \neq 0} (2J+1)$$

6. T. T. S. Kuo and G. E. Brown, Nucl. Phys. A114, 241 (1968).

7. H. A. Mavromatis, Phys. Letters 32B, 256 (1970).

8. R. Mohan, to be published.

9. A. N. James, P. T. Andrews, P. Kirby and B. G. Lowe, Nucl. Phys. A138, 145 (1969).

10. J. W. Negele, Phys. Rev. C1, 1260 (1970).

11. A. Kallio and K. Kolltveit, Nucl. Phys. 53, 87 (1964).

12. J. Blomqvist, Nucl. Phys. A103, 644 (1967).

13. P. Goode, B. J. West and S. Siegel, U. of Rochester, Report UR-875-318(unpublished).

14. J. Blomqvist and T. T. S. Kuo, Phys. Letters, 29B, 544 (1969).

15. J. D. Walecka, Phys. Rev. 126, 653 (1962).

16. V. R. Pandharipande, Phys. Letters 31B, 635 (1970).

17. K. A. Brueckner, M. J. Giannoni and R. J. Lombard Phys. Letters, 31B, 97 (1970).

18. G. E. Brown, and G. Ripka, private communications.

19. W. Gerace and A. M. Green, Nucl. Phys. A113, 631 (1968).

20. P. Federman and L. Zamick, Phys. Rev. 177, 1534 (1969).

21. S. Siegel and L. Zamick, Nucl. Phys. A145, 89 (1969).

22. A. M. Green, Phys. Letters 24B, 384 (1967).

23. A. E. L. Dieperink , H. P. Leenhouts and P. J. Brussard, Nucl. Phys. A116, 556 (1968).

24. M. Kirson, private communication.

25. P. Goode and L. Zamick, Nucl. Phys. A129, 81 (1969).

26. P. Goode, Phys. Rev. Letters 21, 1116 (1969).

27. R. J. Peterson, Phys. Rev. 172, 1098 (1968).

DISCUSSION

HARVEY: One could interpret this last $f_{7/2}$ coupled to the Ca^{38}. Again, it is a weak coupling picture, but we have a different type of core. The wave function, the $d_{3/2}^{-}$ is a particle in a Ca^{38} core, but weak coupling may still be appropriate but just a different core.

KUO: I would like to make the following point about the usage of the experimental single particle energies in a p-h calculation of Ca^{40} or a calculation of Ca^{42}. The experimental single particle energies are supposed to be the poles of the 1-body Green's function, and hence include both the reducible and irreducible one-body diagrams. But in the Green's function formalism for either Ca^{40} or Ca^{42}, the one-body vertex functions should not include the reducible diagrams. Hence, in principle, we should not use the experimental single-particle energies in a calculation for Ca^{40} or Ca^{42}. They should be calculated from evaluating all the irreducible one-body vertex-function diagrams.

CASTEL: How is the problem of the Pauli principle violation avoided in your particle-core formalism?

ZAMICK: In our formalism we do worry about the Pauli principle and we do take it into account - that is why the $(d_{3/2})^{-1}$ times 3^{-} vibrations is not orthogonal to the $(d_{3/2})^{-1}$ times 5^{-} vibrations and that is why we get the coherent effects in our E3 transitions. If we didn't worry about it, if we are completely phenomenological we wouldn't get these coherent effects of alpha alpha prime, so somewhere in our formalism we worry about it. I should say that Greg Seaman has done the decays of these states instead of the alpha alpha prime, the electromagnetic decay. I didn't quite have time to write them down so I apologize to you.

LAWSON: There are now some lifetime measurements by Holland and Lynch on Ca^{41} and a limit on the K^{39} M2. A couple of years ago when Dieter and I looked at the M2 transitions, we found we could fit these by taking an $f_{7/2}$ particle coupled to the core states of $(d_{3/2})^{-2}$. This doesn't necessarily mean that the model we used is correct. However, it isn't clear to me that your model will give these lifetimes correctly. In K^{39} the decay goes from 7/2- to 3/2+ and seems to be only slightly inhibited. In Ca^{41} the 3/2+ to 7/2- decay is inhibited by 60. Both of these are explained by the simple coupling of an $f_{7/2}$ nucleon to the $(d_{3/2})^{-2}$ configuration or the coupling of a $d_{3/2}$ hole to the $(f_{7/2})^2$ configuration.

ZAMICK: We didn't calculate the M2. We can in our formalism--in a phenomenological or macroscopic formalism you can't.

MacDONALD: Larry, this is just a technical point. There are now two known low-lying 2- states in Ca^{40}. The lower state at 6.025-MeV has a large deformed component in the picture of Gerace and Green. Cal Class is investigating the higher state at present. This state may well have an appreciable $(d_{3/2}{}^{-1}, f_{7/2})$ component.

ZAMICK: What energy is that?

MacDONALD: At 6.75 MeV.

ZAMICK: Yes, I think we really haven't tested these wave functions enough, I must agree there are lots of other things that one should calculate like the M2 rates and spectroscopic factors. I think I neglected to mention that if you do a particle hole calculation without the collectivity you will never get the 3- to ground transition rate in Ca^{40} to be large enough. It will be too small by a factor of about 5. That sort of favors the RPA. In fact, the RPA was invented for these 3- vibrations by Gillet to a large extent because of the very strong E3's to the ground.

SANTO: Would you think that the same interpretation as for the low lying 3/2+ state in Ca^{41} would hold for Ca^{43}, because one has a similar state i.e., Ca^{43}, at a similar energy as in Ca^{41}?

ZAMICK: Yes, you can do sort of the same thing. In fact, Goode did Ca^{42} which I didn't bother to write down for lack of time. He did not do Ca^{43}.

SANTO: Ca^{43} is stable. You can do inelastic scattering and check your picture.

IV.B. ALPHA PARTICLE CAPTURE IN THE 2s-1d SHELL*

P. M. Endt
Rijksunversiteit
Utrecht, The Netherlands

I would like to talk about some work on alpha particle capture. Some seven or eight years ago we did some work on that in Utrecht on even-A target nuclei and the gamma rays were detected with sodium iodide counters.

Now I would like to talk about alpha particle capture on odd-A nuclei; in particular, Na^{23} and Al^{27}. The detection is not with sodium iodide, but Ge(Li) counters and the work is done by Manfreued and is perhaps better than the older work.

There are quite a few things rather different if you go over from even-A target nuclei to those like Na^{23} and Al^{27}. One is that the resonance peaks which you would excite if you do this work with a small Van de Graaff generator (ours goes up to at best, 3.3 MeV) are all proton stable, and the Q-value for (α,p) reactions is something like 2 MeV, so the alpha particles have plenty of energy to penetrate the Coulomb barrier. However, if you are looking at states which have a very high spin, and very high would be something like 9/2 or 11/2 or 13/2, then the situation is rather different. The alpha particle emission would be slowed down because of the penetration through the orbital angular momentum barrier and one might have the possibility of a gamma ray transition which can compete with proton emission. Actually, this is exactly what happens, and I will show a few slides. The point then, is, that the only resonances which you see are actually high spin resonances. The low spin resonances just only decay by alphas, if you look at least at higher energy radiation. And of course, these high spin resonances then in turn would decay to high spin ground states so this is a very selective sort of reaction to look at rather high spin states. Actually you excite the states that you just don't see in any other reaction. At least at excitation energies of say, in Al^{27}, from five up to something like 8 MeV, we find quite a few new states which would not be as excited say as $(He^3, d\gamma)$ or (p,γ) or something like that.

Might I have the first slide? We could start with this one. This is for the Al^{27} $(\alpha,\gamma)P^{31}$. There are some pronounced resonances starting at rather high energies. Let's say this is 2.9 MeV and it goes up to 3.3 MeV. At present, with the spectra shielding system we have a rather low background and these resonances stand out much nicer, also a very few weak ones between 2.5 MeV and 2.6 MeV, but they are just too weak to do any significant spectroscopic work on.

---

*Transcribed from the tape recording of Professor Endt's talk.

This is the region for gamma ray energies from 7 to 13.5 MeV and again it covers, in this case, from very low energy to about 2 MeV and then it starts at 2.4 MeV again up to 3.3 MeV. The yield is a little higher for the Na^{23} (α,γ) Al^{27} and the resonances are much nicer. Here, at least one has something like 8 or 10 resonances on which one really can work. They are still plenty weak, I mean (α,γ) is not a reaction in which you really have very strong resonances, but with a big germanium counter, we have a 60cc one, that would do quite some work, I would say. This is the lower energy particle of 3 to 5.5 MeV, and these two are (α,p). This is the $(\alpha,p_1\gamma)$, so one detects the gamma rays from the first excited state in Mg^{26} to the ground and the point is that these resonances just have no connection with the resonances which one sees here. For instance, this is rather small part, and has quite a few pronounced and a row of small resonances here which one doesn't see, and the other way around. These are then the low spin resonances and these are the high spin resonances. This part is the gamma ray transitions from the second to the first excited state in Mg^{26} and that only starts at something like 2.6 MeV or so, so this is both (α,p) and (α,γ).

So far, for the change going over from even-A targets to an odd-A target--there is also quite a change if you go over from the sodium iodide to the germanium. In the sodium iodide work we had the nasty experience that it is very difficult to see anything of the low energy gamma radiation. The part up to something like 7 MeV in the spectrum is rather completely obscured by capture gammas for neutrons are produced by (α,n) reactions. So there are so many neutrons around they mess up the spectrum at something like 7 MeV, so the other transitions that we can work on normally with the sodium iodide are those going to the ground state or the first excited state or, if it is a strong resonance, maybe to the second excited state, but that is about all. You see very little or nothing about the decay of those excited states which you excite. It is quite different with the germanium detector. The cross section for the neutron detection is smaller and the background just has more or less disappeared. The counter is smaller of course, that helps also, so might I have the next slide which shows a spectrum. Lots of gamma rays, even down to something like 1.8 MeV. This is, again, the gamma ray from $(\alpha,p_1\gamma)$ so below that you have a very intense count on which everything else would sit. But above that one has a rather clear spectrum and there is quite a lot more than you can do with the sodium iodide I would say. These would be transitions which go to the 3 MeV state which is 9/2+.
This is a transition, these three peaks here, going to the 4.51 MeV and 11/2+, so already just a brief glance at the spectrum like that shows you that you probably have a high spin resonance which decays at least to two known high spin

states and then to quite a few unknown states of which the three we have here in the 7 to 8 MeV region have very high spins.

The next slide shows another spectrum, but now taken with a calibration spectrum between, I will tell a little more in detail how one got this. It is rather nice idea which Van der Leun had, who calibrated germanium spectra with the higher energy particles. It is easy to calibrate with germanium spectra up to something like 3 1/2 MeV or so with radioactive sources. From there on, there was not very much available. People had been calibrating on the F^{19} $(p,\alpha\gamma)$ O^{16}, 6.13 MeV gamma rays, but while this is alpha particle work so you would not have that, and there are also troubles connected with that. It is not a very good one to calibrate not to go into too much detail, but anyway, what I can do is to and which is rather a general matter to calibrate the higher energy part of a germanium spectrum, that is to use a small calibration source plus beryllium source [a Po-Be source] to produce alpha particles and the neutrons in the beryllium and the neutrons are slowed down, you put the whole source in a pail of water and we have a little source which costs about a thousand dollars and it provides one some 2×10^6/sec and that is plenty to have good calibration spectra in material which you would put in a ring around your germanium. In this case it was iron but you can use copper or nickel or aluminum or quite a few more. Anything containing chlorine would be fine. Chlorine has something like a 35 barns cross section. All these factors can be used to calibrate the germanium for high energy particles especially these lines in the iron are very well known. They are the two lines which lead to the ground state and the first excited state in Fe^{57}. They differ by 13 keV, the famous Mossbauer level and the energies are very well known. They are wonderful for an internal calibration. The idea is of course that you do it simultaneously and there are no troubles with gain shifts or things like that.

May I have the next slide. This shows it's a little old and at present we would have a few more states, but at least some six of them I think from here are new states; also the spins and parities are still omitted from a little later to that. These are the resonances and when we see for instance that here we have this 9/2 state that is reached in very many of the resonances and also very many of the higher bound states decay to the 9/2+ state. This is the 11/2+ state and again, many resonances decay to that one, and at least one of the higher bounds also go to the 11/2+. So it is plainly high spin states which one reaches here.

Next slide, please. Oh, I forgot to say in some way all the states which we see seem to have even parity. I will come to that a little later but anyway, we have no indication of any odd parity resonances or bound states which we excite in this particular case. Here is more or less the same thing in $Al^{27}(\alpha,\gamma)P^{31}$, again we excite many unknown states and this is up to something like 10 MeV, and again, for most of the resonances and the bound states which we excite, whenever we have been able to determine the parity it turned out positive. One exception is the resonance at 2.9 MeV, which is the one which decays in this funny way to all the higher excited states and that is known to be an odd parity state by the (α,p) work. It is a weak resonance in (α,p) but one which decays by (α,γ) and (α,p) in comparable strength, so the spin and parity could be obtained from the (α,p) work and we think that these are the positive parity states. It is also the only one for instance which decays to the known 7/2- state at 4.33 MeV in P^{31} whereas most of the other resonances go for instance to the known 7/2+ at 3.41 MeV. They decay strongly to the known even parity states; and there is only this one as far as we can see, which goes to odd parity states.

To get at spins and parities one would do to start off with angular distribution work. Normally, angular distributions alone are not sufficient to determine spins and parities. But we can combine this with Doppler shift work on the lower of the secondary gamma rays and so get a lifetime of most of these states and that together proves to be sufficient to determine spins and parities in an unambiguous way for most of the states. Actually, it is much the same way as it is for (p,γ) work. We would no longer do difficult gamma angular correlation work, but mostly just angular distributions together with lifetimes. Hopefully to gather data from something like (He^3-d), that is also a proton capture reaction, which would supply the parity but not the spin and the spins one can then get mostly at least in an unambiguous way from (p,γ) plus lifetimes. Actually, we haven't done any angular correlation work for the last three years or so, I would say, and it is fine. The analysis is rather difficult, it takes a long time and this is just so much easier. The angular distributions have one funny feature and that is over at least from different from (α,γ) on an even-A target and that is that you can populate, well, let's only stick to $Na^{23}(\alpha,\gamma)Al^{27}$ on which we did most of our work. You can populate two magnetic substates 1/2 and 3/2. So the angular distributions would have an unknown ratio of the population of these two, but in addition to the mixing ratio of the gamma rays. If you are looking at the angular distribution of the primary gamma ray you would have two parameters to determine from the experiment, that is , the ratio of the two population probabilities for 3/2 and 1/2 and the mixing ratio. However, it turns out that the angular distribution is extremely insensitive to this population ratio. It is because mostly the orbital angular momenta of the captured particles are quite large and this means that the spin of the resonance is directed nearly perpendicular to the incoming beam and that the angular distribution is insensitive whether you take 1/2 or 3/2 or a mixture of the two. That of course makes it very much easier. You can just take anything. Say you take the ratio equal to unity or something like that and then you just analyze from that. I will show one slide which one can see that the difference is really very, very small and it makes it just so much easier. Now there is only one more parameter left and that is the mixing ratio of the gamma ray in question.

Might I have my next slide. It shows some of these angular distributions. This is from a state at 12.38 MeV and let's first look at the transitions going to known states to the 3.00 MeV 9/2+ state for instance. From the chi-squared plot which we have here we see that the 9/2+ possibility for the resonance is eliminated at the 0.1 probability limit. The choices to 7/2 or 11/2 give very nice sharp minima and mixing ratio of 0. The fits can be seen here. This is the best fit for the 9/2 and this for the 11/2 or 7/2. There remain two possibilities for the spin of the resonance, now there is also a transition to this 4.51 MeV 11/2+ state and if we do that we see the 11/2 is here, the 9/2, which is already out, is here, and the 7/2 should be pure quadrupole and it is up here. This rejects the 7/2 and we only have left the 11/2 possibility for the spin of the resonance. Again, we have here the peak which nicely rejects the 7/2. Some of these go from four thousand counts up to 7,500, so these are quite nicely anisotropic some of them. That is clear of course because you have the spin very much anisotropically directed, the spin of the compound state. Once we know the spin of the resonance this same resonance can be used to determine the spin of the lower states. Here we have, say the 5.67 MeV. These are possibilities. These are 7/2, again pure quadrupole and here we have the 9/2 possibility so it can only be 9/2 and again the fit is rather nice. Another state at 7.1 MeV, that is one of these rather high states. Again, 9/2 is the only possibility, as 7/2 is rejected rather like the preceding one. Again, with very little mixing shows almost zero in the error.

Next slide please. This is more or less the same thing for the resonances. It works exactly the same way. Here we have already a unique determination only from looking at the transition to the 9/2+. There is only the 11/2 which has a chi-squared below the .1 percent limit. That is the 11/2. Once you have the 11/2 we find that the 5.3 MeV level has spin 9/2. It seems a very straightforward and simple sort of game to play. Altogether I have not taken into account the results can be more or less, if we take them together, there are branchings rather accurate excitation energies for these new states, mean-lifes from Doppler shifts and alpha gammas wonderful for Doppler shift and of course have been quite large as compared to our (p,γ) work, your recoil damages would be quite a little bit larger and some unique J-values for 6 bound levels in Al^{27} and for some fine resonances and some unique parities, all even parities for some, I think, 8 states. So far for the (α,γ).

This is one way to gather spectroscopic information on nuclei in the sd-shell. What can we do with it? One way of course is to do shell model calculations Oak Ridge and Rochester have taught us how to do it. But what one would like to do of course is at least on the even parity states is to use the full 2s-1d shell, just any number of particles in the three subshells in the of the 2s-1d shell, I have just figured out that would take you up to a maximum size of matrices in the 3+ states of Al^{28} of 6,706, so that is the very worst that you can have. You would have to mix together some 7,000 states. It is still less than I thought previously. I heard people talking about many 10,000 or something like that. Somebody even said a 100,000 but they never counted, but if you really counted, it is only 6,700. I would say that maybe after all there might be sometime some hope that we really cover the whole 2s-1d shell. I don't know how fast computer development will go, but there might come a time when we really do this whole shell. At present of course there is no hope of diagonalizing matrices of 7,000 x 7,000 and we rather have to be content with somethink like a few hundred then you could go from something like 16 to something like you have done in Ne^{22} and start it on A = 23, I think here, but then you already go up to A = 22, you would go some 500 x 500 to 900 x 900, something like that, very close to a thousand at least. Beyond that it is still hopeless and one has to make some sort of restriction. I think you look for the up to 28 the restriction was made that there are no particles in the $d_{3/2}$ shell that is one possibility. From 28 up to, I say, 35, one can make the restriction that you only have a limited number of holes in the $d_{5/2}$ shell so that goes down to up to 2 holes in the $d_{5/2}$ shell and that is from 35 on you again work in your unlimited 2s-1d shell space, just taking holes instead of particles. We have gone a little further with the region from A = 30 to 34 in the initial calculations, Oak Ridge together with others from Utrecht, calculate energy levels, excitation energies, stripping and pickup spectroscopic factors and as a sort of by-product you have the wave functions which of course you can use for a lot of other things, like calculating M1 and E2 transition probabilities, ft values for beta decay, static moments, of course, dipole and quadrupole moments. This is what we started on this 30 to 34 region actually because it has the most abundant experimental information at the time. There are some 40 M1 transitions with known transition probabilities, some 40 E2's; some 25 ft's; 3 magnetic dipole moments and one quadrupole moment, so one can try to fit those. Let's see what kind of para-

meters can go into such a kind of calculation. The energy levels were obtained with the modified surface delta interaction so that would require 4 parameters. It is a very easy sort of interaction and you would have the three binding energies of $d_{5/2}$ and $s_{1/2}$ and a $d_{3/2}$, so that is seven parameters altogether which you would fit and they are fitted from the whole region of 28 up to 30 or 35 or something like that. Let's not bother about that. They just add to the wave functions and you take that for granted. Wave functions we don't question anymore. Then to calculate transition probabilities you can use a few more parameters. Let's first take the E2's for instance, if you would just do that with no effective charges, you would find that all the E2 are underestimated by the theory so you take some sort of effective charges, so that would be two parameters, one for the proton and one for the neutron. In fitting these 40 E2's we find the best values, the proton charge is 1.44e and the neutron charge is 0.68e. So the neutron charge is a little larger than the proton charge but on the whole they are, at the average about .5 which so many people take in this respect. Of course you would expect that these effective charges depend on the size of the configuration space which you are taking. Glaudemans has a nice little calculation. Starting off with a very small space and then taking some a little larger and finally as far as we can go and this was on I think S^{33} there is a slide on that.
This is to show what kind of agreement you would get on the excitation energies. This is P^{31}. This is the experimental part and these are calculated energies and on the whole they do not agree too badly. At least the number of states which we have is the same in both slides, without omitting. I omit of course the odd parity states. It is only the even parity states which we use, and not all spins and parities are known. There would be two more known at present, what we think they are at least. The average deviation would be of the order of something like 200, maybe 250 kilovolts or so. Next slide please. This is the one I was just talking about. This is the quadrupole moment of S^{33} the E2 strength in Weisskopf units for the first excited state to ground in S^{33} and the second excited state to ground. In your first calculation you will just assume that the ground state is 1 $d_{3/2}$ particle outside a closed core. Here you would assume for the first excited state 1/2+ state that it would be $d_{3/2}$ squared and a hole in the $s_{1/2}$ shell and this is a five particle state which you would assume as being $d_{3/2}$ squared and the hole in the $d_{5/2}$ shell. The average charge for proton and neutron would be 1.3. This is some intermediate region. I don't quite know how large I took it, but it amounts to diagonalizing some 40 states or something like that and that average charge turned out to be .7 and this is finally the space in which we have done all these

calculations with up to 2 holes in the $d_{5/2}$ shell and now the average charge is .5 and the agreement with experiment is not bad.

The E2's then are just slowed down the best values which you would have if you fit the known transition probabilities one can say if you just take the strong E2 transition we have taken 14 strongest. We would like very much to have good agreement for the strongest percentage-wise, whereas for the weak ones you never can expect to get a very good agreement, again percentage-wise. You are just very happy if the weak transition turns out to be weak but of course the transition probability would be very sensitive to any change in the parameters you would have had for a weak one because it is normally that large components cancel, so let's not worry too much about weak ones. I think the weak ones turn out to be weak but for the strong ones we would have an average of 9.1 by computers and then the variation between the theory and experiment the average delineation is in this case 3.5 or that is something like 14% and agreement for the strong one. Just a little MeV variation is just a little higher experimental error on these strong ones. They are obtained you have to know three experimental pieces of the information to get the E2's the lifetime, the branching and, finally, for most of them the mixing ratio to arrive at the E2 stage. The errors on all three enter into the final value of the transition probability. In looking at the M1 again, we have something like 40 M1's if then one would just calculate let's say realistic or if you like bare nucleon g factors you would have all this on the average you would arrive at far too strong M1 transitions calculated. What one does one quenches the g factors, uses effective g factors and again, one can treat these g factors as parameters. There would be four of them for the proton and neutron for the orbital momentum, orbital angular momentum part for the spin part so that four parameters would be used in the fitting.

If one tries to do that, one finds that it is just not very well possible to fix the whole region from 30 up to 34 with just these four parameters. That is more or less understandable because one expects the lower part to be worse and there you have this influence of the restriction that we took up to two holes in the $d_{5/2}$ shell, but the closer you work to the $d_{5/2}$ shell, the worse the agreement would be one can expect. In other words, the larger these quenching factors have to be, so we split up the region in two, 31 and 32 in one group and 33 and 34 in the other and then these four g factors were fitted separately for these two regions. If we do that, and again we look only at two of the stronger ones, these states would be 9 of the stronger and one the average strength from 19 Weisskopf units and the average delineation between theory and experiment is .08 Weisskopf units just about the same as we would have for the effective charges. That is the second region. That is for both regions we cannot come out of it. This is both taken together, three different sets of parameters which arrive at about the same sort of variation for each. There is another way to do this however, and that is not to take the g factors as parameters but rather let's say the speed of the elementary transitions. One could say that in elementary physics the transition would be that for instance you have the M1 operator causing transitions of the form $s_{1/2} \rightarrow s_{1/2}$, $d_{3/2} \rightarrow d_{3/2}$, $d_{5/2} \rightarrow d_{5/2}$ and $d_{3/2} \rightarrow d_{5/2}$ or visa versa. You could even go a little farther and say going from $s_{1/2}$ to $d_{3/2}$ that strictly would be forbidden for the M1, but one could still consider it as a variable parameter which you might try to fit. Of course $s_{1/2} \rightarrow d_{5/2}$ is also forbidden so one should take that into account. If one does that, then we have four of these elementary transitions and one can split them up into an isospin part and an isovector part and that would then amount to 8 parameters, which one can again hopefully fit from this set of forty M1's which we have. It turns out however, that quite a few of these 8 parameters come in very little to the calculation. They are not very important. Rather, what was done then is to start off with a fitting all 8 of them and finally keeping I think five of them fixed and only for a further fitting, three most important ones and they are the isovector part of the $s_{1/2}$ to $s_{1/2}$ and the $d_{3/2}$ to $d_{3/2}$. The $d_{5/2}$ comes in very little actually. It does come in very little to the weak wave functions. There are wave functions which are 50 percent $d_{5/2}$ at the whole matrix. If one does this there would be three parameters to be compared with the four parameter which we have for the fit with the effective g factors. Again, we would have the average of .9 Weisskopf units, but now the deviation between the theory and experiment has gone down to .04 Weisskopf units which are just a little more than 20 percent. That is about equal to the experimental errors I would say. Especially, if you take into account there is still not a very good procedure of indicating errors and lifetimes which have been obtained from Doppler shift measurements and slowing down and should go into this and some people take them 10% or 15%, or at least the uncertainty of most of these lifetimes in only slowing down I would guess is something like 15%. One would say that more or less for these transitions that there is agreement within the experimental error. One can also do the various ft values and the average variation becomes for the 15 fastest transitions, Gammow-Teller transitions, the variation would be the longest ft value .1 which is something like 33%. Again, you would not have good agreement, at least for the very slow major transitions, for instance, there is S^{32} right in the middle of our region with an ft of 7.9 and I think the best we can get it is 6.2 which is still a factor of 50. I don't think that is very important. Just by changing the parameters a little it would change the value like that very much so that the weak transitions you should be very glad if you obtain them weak. What one can also do is compute E2 over M1 mixing ratios. There is a difficulty in this and that is each different group would work with different sign conventions so if you really want to have a meaningful comparison for example, the signs which you compute and which are measured you would have to reduce all those to one sign convention. This has been done and we have now in this region, 17 good mixing ratios which are significantly different from zero and they all have the correct [same] sign. We think we are good at least in the signs of the mixing ratios.

In one particular transition you had to have a lot of different components. I mean you would start off with the wave functions which has come to some 300 components and your final state would have some 300 components so altogether there would be something like what it is 10^6 contributions to a particular transition. Of course, many are forbidden for some reason or other. Still, let's say you would have a few ten thousand easily. We don't ask the computer to write out all of them, but only let's say a few hundred of the largest, so you certainly mix only above that limit. There is some rather nice fittings in these numbers. At first sight you have the idea that you cannot do anything with them, but one can, for instance, in looking at the E2 you find that there is no large contribution to any particular transition, for instance to the strongest E2, the strongest component would only contain something like 10%. So actually, an E2 is made up of a contribution of very many of these elementary transitions between one configuration and the initial and one in the final state. I want to say it another way. That E2's are very collective. I mean if everybody knew that E2's were collective, it is nice to see it confirmed in this sort of large shell model. On the other hand, the M1's are quite different to look at. You find at least for the strong M1's just one contribution which does about everything. You will find that it contributes more than a 100 percent because the other parts might have a different sign so you might have one contribution and then you would have to have 120 percent or so. But so the M1's are very much single particle. The strong M1's are at best really you could do a lot of them, but I have just taken the easiest configuration. The most important configuration would give you a very good estimate of any strong M1's which is nice for people doing let's say calculations on analog to antianalog states or things like that. Again, if you then analyze which components are most important you find that all the strong M1's in this particular region are caused by the large isovector in the $s_{1/2}$ to $s_{1/2}$ transition, so the M1 is operating on the $s_{1/2}$. And this is more or less in agreement with what we derived for the analog to antianalog transitions. The analog to anti-analog transitions are quite large if the M1 operator operates on a particle in an orbit which has j = 1 + 1/2. If you operate on an $s_{1/2}$ or $f_{7/2}$ or $p_{3/2}$ orbit, then you get strong transitions, but you get very weak ones for let's say the M1 operating on the $d_{3/2}$. The reason is very easy, I mean just for if you take the orbital momentum and the spin parts you find that for the j = 1 + 1/2 they are in phase and otherwise they are out of phase and you get nothing strong, something the order of a factor of 100 or so lower transition probability. Again coming back here we would have an isovector part and an isospin part but normally Maripuu's rule will also hold for any M1 transition even if there is an isospin part because it is small anyway. That is what occurs here for the $s_{1/2}$ to $s_{1/2}$ you find strong transitions but for the $d_{3/2}$ to $d_{3/2}$ rather weak ones. As I said the $d_{5/2}$ components come in very many small parts and altogether we do not have in this region at least very strong, single particle $d_{5/2}$ to $d_{5/2}$ transitions. There are also some groups of states for which we do not find a good agreement. The first group which we have no explanation for, are the 1+ states which are very, very bad. I have no idea why, but somehow off by a factor of 10 or so. There is something essentially wrong with 1+ states, but I don't know why. Another one and that is much easier to understand, another group, and that is why for these the spectroscopic factor is a predicted spectroscopic factor which is in strong disagreement and that is why you predict a strong spectroscopic factor and experimentally you find a very small one for either pickup or stripping then that state you better leave it out and you can expect to find a good agreement. There might be something wrong with the J value or something like that, but anyway, they don't fit. I have two more slides which are not quite concerned with what I have been telling about which I would like to show. Every piece of experimental transition checks within the experimental error in the calculation. It is the lifetimes, 1200 experimentally, and 160 predicted, 175 seconds and one 55,200, 163, the spectroscopic factors .69 .61 .27 .28, weak, .002 predicted .005 and .008, the branchings are predicted well and the mixings, this one for instance is 1+, 0.75 predicted plus 0.65 so that is just fine. These are just MeV's. The average parameters which we derive for the whole region. The next slide is Si^{31}, it is only one piece of information on this one which is not quite all right. The lifetime of the first excited state 955 seconds measured and 2000 predicted. Here is an old slide and the lifetime I think has gone up a little and the average we have now would be a little longer. The prediction has gone down a little, but they still do not agree. This is still one piece of information where the information is out with the measurement. All the other things again are good as one can have. Next slide please. That is P^{32} and that is the case where we have the half way 1+ state, the ground state is 1+ and let's say for instance the lifetime of the second excited state is wrong by a factor of 10, but the transition from the 0+ to the 1+ and a lot more things connected with the one plus states are not all right. But on the other states again, it is mostly as good as you can wish it to be. All right, that is the one that I wanted to show. These are the predicted quadrupole moments of the lowest 2+ states in even-even nuclei on just the simplest shell model which you can take, so that is much simpler than the large space which we have just been taking. It shows that the predicted values go from negative to positive and in S^{38} and the negative is finally positive. Our measured values are here and they check not too well it is without effective charges if you increase the effective charge you also get the quadrupole moments. This one is not bad and there is now a new value for S^{32} which is negative in beam and which is more fitted here. I think that it is a little too large. So this funny behavior is starting off negative and then going positive with the negative again. That part is all right. One would very much like to have the S^{36} first excited state quadrupole moment to see if it really turns positive again. Oh yes, and then the last slide is also something which came up just a few days before I left. It shows the ratio of mixing ratios E2 over M1 for corresponding transitions and types in the 2s-1d shell. One can say that one has this mixing ratio which the E2/M1, the ratio has large isoscalar parts plus a small isovector part, so the next problem is isoscalar which comes from an isovector and over the amptitude for the M1 there is just the first and the isoscalar part would be small and the isosvector part would be large. We can just about forget small parts.

It became clearer -- if you take the mirror transitions then these would change signs where A would remain the same so the spectra has the same magnitude, but opposite signs for the mirror transitions and that is just as is shown here, and the ratio of the two so it should be a 1- or they put in a minus sign already, so it is for something like I think 20 transitions in them and for 18 of these, they are very nicely on this line of unity. There are two which are off, but if you look at the data then the experimental data you find something fishy with both of these measurements. I don't quite remember in which nuclei they were. There are a few more, this one I think is also opposite. If you take the average you find that you would find the chi-squared which is just about unity and normalized chi-squared in this group, so that seems to be fulfilled very well, and one has to trust in the simple predictions like this.

DISCUSSION

WARBURTON: I wondered why there weren't any states with

spins greater than 11/2 showing up in the two (α,γ) reactions you did?

ENDT: I don't know. We have 11/2 resonances. We might excite 13/2 states. Maybe they are rather weakly excited. We don't have the spins and parities of all our new Al^{27} states so it just might be that they are not among the group for which we could get a unique spin and parity determination. They are not very strongly excited.

WARBURTON: One other question. What was your philosophy in including in the least squares fit for M1 transitions a term which you know from first principles must be zero; namely, the $2s_{1/2}$ to $1d_{3/2}$ transition rate? I didn't understand that.

ENDT: It might very well have an effective value which is non-zero. That they are effective comes about because you cut our part of your shell model contribution. Like say the lower part of the $d_{5/2}$ shell, or the p shell or the fp shell, and they might work together to give you a non-zero matrix element. For the A = 30 to 32 region the best value is 0.4 ± 0.1 in whatever units it is. So that is non-zero of the $2s_{1/2}$ to $1d_{3/2}$ isovector M1 matrix element.

BAKHUR: In both your (α,γ) spectra I saw even-parity states only. What happened to the odd-parity states?

ENDT: I just don't know. It just must be that the M1's are so much stronger than the E1's. The question remains, why do we excite only even-parity resonances? That I don't know. They all seem to be 1 → 4 resonances, and no 1 → 3 although that would also be a rather nice high value.

BAKHRU: And another technical short question -- how do you normalize actually an angular distribution? What was your normalization?

ENDT: We have one counter sitting at 90° which acts as a monitor, in this particular case it was sodium iodide.

SANTO: You mentioned that you had troubles with the 1+ states, that the wave functions didn't fit your data. I think for the transition probability that is also systematic for two particle stripping in the (He^3,p) reaction. You also have troubles getting the right magnitude of the spectroscopic factor for the 1+ states and only for the 1+ states. This was found for the $f_{7/2}$ shell, such that it cannot be a feature of the s,d shell alone.

ENDT: You mean if you do simple shell model calculations, you get different spectroscopic factors from what you really measure?

SANTO: Yes.

HOROSHKO: With regard to transitions to 1+ states in the odd-odd nuclei, we find something funny also in the (d,α) transitions to the 1+ states in Cl^{34}. We analyzed the alpha data using the earlier wave functions of Glaudemans and find relatively good agreement for the relative cross sections and L mixing for transitions to the first 3+, first 2+ state and also the second 1+ state, but not the first 1+ state. My specific question is: do you find significant $d_{5/2}$ admixture in the 1+ states?

ENDT: I just don't know it by heart. I couldn't say. I don't have the wave functions here. I could look it up.

PROSSER: First a comment. We have some new mixing ratios and lifetimes in Cl^{33}. Dr. Glaudemans was kind enough to supply us with his predictions for Cl^{33} and in general they agree about as well as the S^{33} fits we say on the slide. And next a question. Has Dr. Glaudemans revised his earlier fit and included any of our data, which I believe we sent him?

ENDT: I don't think we have. We stopped including any more experimental data around March of this year.

TITTERTON (Chairman): Would any of the theoreticians like to comment on this 1+ difficulty?

HARVEY: I think there is trouble also in, what is it, Na^{22}. In that case you can see that this is particularly sensitive to the spin dependent part of the force, the two-body spin vector and tensor. Now you haven't put these in explicitly but since you have truncated your space they have sort of crept in the back door. But they still may not be strong enough. You may be sort of measuring the lack of spin dependence. Do you know how many of the energy levels of the odd-odd nuclei when into your fits, or did you only fit on the even-even states?

ENDT: No, no most of the others went in. We threw out the 1+ but most of the others are taken along.

HARVEY: Yes, to determine the parameters of the interaction.

ENDT: Then I think the 1+ states are also taken along.

HALBERT: Yes, they were. But the Hamiltonian form that was assumed, (the modified surface delta), although it has seven parameters, is quite "stiff" and may not allow you to get in parts simulating two-body spin-orbit and terms of contributions. What I don't remember is whether the resulting shell model spectrum did fit the observed 1+ positions closely.

BRUSSAARD: I might make a comment on Dr. Warburton's question concerning forbidden $2s_{1/2}$-$1d_{3/2}$ M1 transitions. One can easily write down a diagram that shows that there can be a contribution in higher orders of the residual nucleon-nucleon interaction. In zeroth order process (a) is ℓ-forbidden, of course, but one can consider process (b), which is of first order in the residual interaction V. For a $2s_{1/2}$-$1d_{3/2}$ shell model calculation p can be a $1d_{3/2}$ particle and h a $1d_{5/2}$ hole. If the complete s,d shell is taken as configuration space, one cannot have $1d_{5/2}$ holes, but then the ph bubble in process (b) can be, e.g., a $2p_{1/2}$ particle and a $1p_{1/2}$ hole. The latter possibility will contribute less, but it shows anyway that there can be a finite contribution. And that is enough to justify the introduction of an effective M1 moment for $2s_{1/2}$-$1d_{3/2}$ transitions.

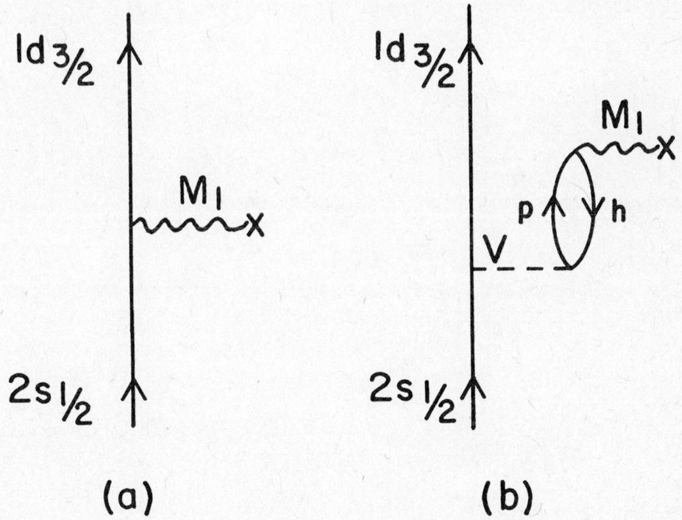

(a) (b)

V.A. AN APPROACH TO SOFTWARE MODULARITY IN DATA ACQUISITION

Rudy Penczer and Richard Hully
Digital Equipment Corporation
Maynard, Massachusetts
(Presented by Rudy Penczer)

I. INTRODUCTION

One of the most vital considerations in the design of a general purpose data acquisition system is its adaptability to a specific experiment. The average set-up time for a series of related experiments will be significantly reduced if the same general on-line system can be used for each - with only minor, rapidly accomplished modifications. A logical approach to the solution of this problem is the development of a general data acquisition system in a modular form.

The concept of modularity extends to both the hardware and software "building blocks" of the system. These components or modules may then be combined as they are or they may be individually altered to match a particular experiment with specific requirements - resulting in a reduction in time and money. The best system is one which closely approximates the needs of a typical experiment, while remaining easily modifiable for other different tasks.

Altering computerized systems requires a detailed knowledge of hardware and software modules. This paper will discuss a system composed of such modules. In particular, the software modules will be discussed in detail with emphasis on both their functions and on their structure. The example chosen is a highly modular pulse height analysis data acquisition system utilizing a PDP-15 computer.

II. AN OVERVIEW OF THE COMPONENTS OF A MODULAR DATA ACQUISITION SYSTEM

It is convenient to divide the system discussed in this paper into seven main components. Of the seven, three are hardware and four are software.

A. Hardware

Computer
Peripherals
Experimental Equipment

The *computer* is made up of three main units; memory, central processor (CPU), and Input/Output (I/O processor).

The memory should be expandable, the maximum expansion being determined by the probable experimental requirements. The I/O processor has *associated* with it a bus system by which peripherals can be added as required. It also has the facilities to handle each device at an appropriate speed.

Many different *peripheral devices* may be necessary at various stages in an experiment. Here are four examples of commonly used peripherals and their functions:

> A mass storage device to provide fast access to and high speed transfer of a non-core resident routine (e.g., a disk with 60 KHz transfer rate and an average access of 17 msec.).
>
> A mass storage device to provide medium speed transfer and access to non-core resident routines or data, and which can be removed from the system for semi-permanent storage. (e.g., DECtape - a non standard magnetic tape system, with a transfer rate of 5 KHz.)
>
> A simple low speed device used to load boot bootstraps and also the simplest device in the event of the need for system maintenance. (e.g., paper tape with a read rate of 100 words/sec.)
>
> A device to provide communication with a computer center. Reliability and compatibility are the controlling factors, not speed.

The *experimental front-end equipment* will of course be specially tailored to the incoming data. It includes the hardware necessary to accept experimental data and present it to the computer. The most probable element of any nuclear data acquisition system will be a set of nuclear ADC's with a very high transfer rate - 40 KHz burst rate. Additionally, a variety of other experimental equipment is used (e.g., scalers, voltmeters, etc.). Most of this equipment does not require the very fast service that an ADC requires. The priorities for servicing these devices and the speed at which data is transferred from them should be chosen by the used. One approach to a hardware frontend that is gaining in acceptance is a standard CAMAC I/O system (see Figure 1). This concept is an excellent example of the value of hardware modularity in cutting down experimental set-up times.

B. Software

In addition to the three hardware components above there are four main software components:

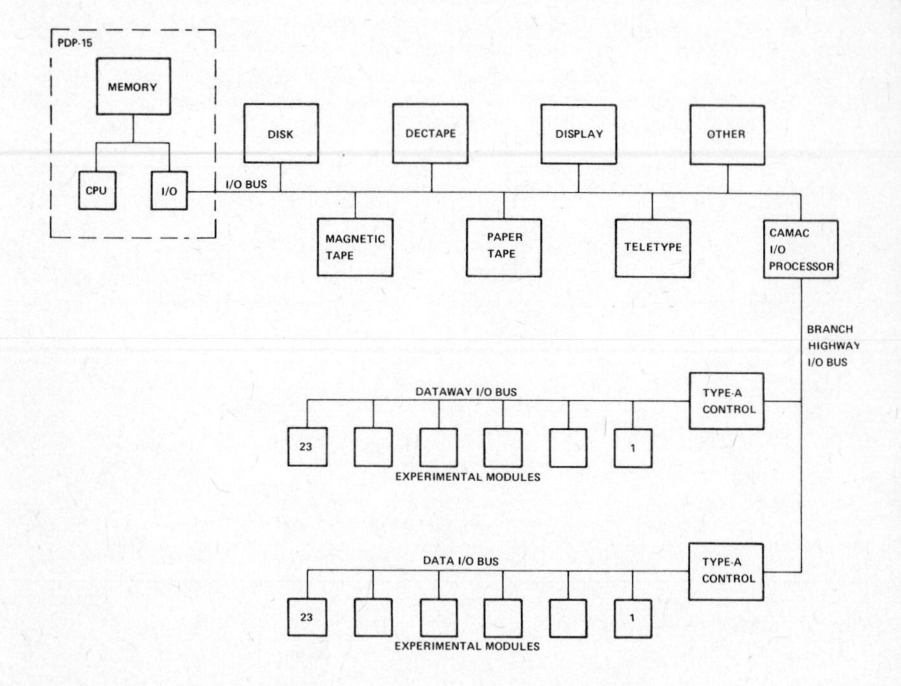

Fig. 1. The PDP-15 I/O System.

Monitor
Experimental Device Handlers
Main Program
Systems Software

The *monitor* is central to a modular computer based system. It serves as the nerve center for the control of all communication between all modules both hardware and software, for the building of new software units, and for the operation of the whole system. It is the framework into which all other software modules plug.

The *Experimental Device Handlers* handle the incoming data and permit the devices to operate in all required modes; (e.g., two ADC's may operate as either two independent units in a single parameter mode or as a single unit in dual parameter).

The *Main Program* receives and executes user commands. There will be differences in the requirements of user commands to be executed (e.g., single or dual parameter); and therefore the main program must be capable of inserting a variety of software modules to accommodate these and other user options in the experiment.

The *system software* provides the user with programs that allow him to build, test, and run his system simply and efficiently. (Figure 2).

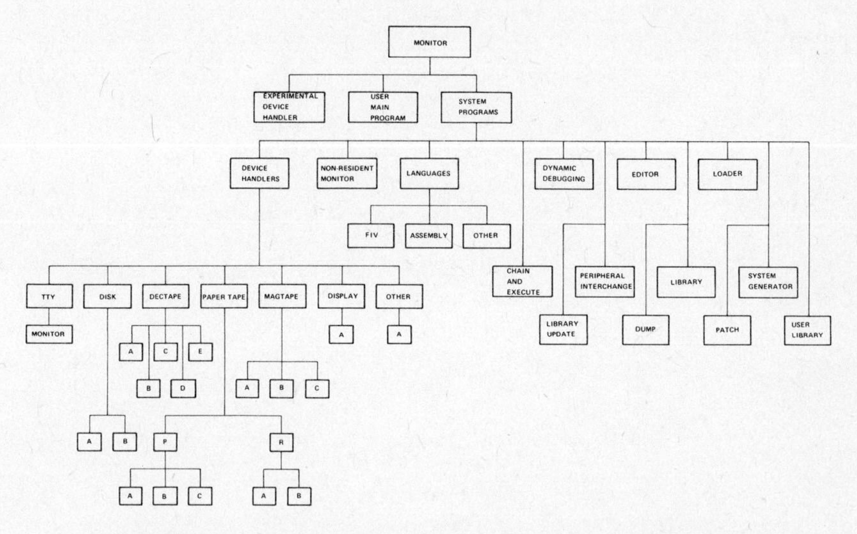

Fig. 2. The PDP-15 Monitor Software System.

To utilize a modular data acquisition system the user must understand the above seven (7) components in detail. A variety of excellent information sources are available from the computer manufacturer on hardware components and on the systems software. This is not true of the remaining three software items. This discussion will, therefore, concentrate on monitors, experimental device handlers, and the user's main program.

III. SOFTWARE COMPONENTS IN DETAIL

A. The Monitor

The monitor acts acts as the "nerve center" for the creation, addition, and utilization of new routines as well as the modification and uitlization of existing routines. It is the "software rack" into which other software modules plug, and as such it controls each module in the same generalized fashion.

The monitor consists of two parts: a non-resident portion, which is used in system set-up, and a resident portion, which is used in system operation. The non-resident portion can be considered in the same category as a compiler in that it is a system program and is loaded by the resident monitor when required.

The monitor has eight functions.:

1. The monitor is the framework for the operation of relocatably coded software, (e.g., the software modules can be plugged into any available space in core memory).

2. The monitor provides an input/output (I/O) programming system for the user controlling I/O in a *device independent* environment. The same commands are used for a variety of devices. Device independence is discussed in more detail in the Appendix. Within the I/O system, the Monitor performs the following:

a. It must keep track of which functions are to be performed by which device. A table is kept correlating physical devices and their logical assignments, resulting in a device independent environment.

b. The monitor uses those device handlers required and maintains a table by means of which the user program can access the handlers.

c. The monitor actually controls the functioning of the I/O by determining which device requires service and using the above tables, causes the proper service routine to be accessed.

d. The monitor has an I/O system error diagnostic routine that will inform the user of the nature of any occurring error; e.g., illegal data mode, illegal function, too many files referenced simultaneously, etc.

3. The monitor communicates with the user via the devicehandler for the user input device which is contained in the resident monitor. The teletype is an ideal device for this purpose.

   There are overriding control functions that the user may want to execute during system operation:

   An unconditional return to the monitor.
   A dump of the contents of core to mass storage.
   A restart of the user program.
   Deletion of a command prior to its execution.
   Termination of a long message; e.g., the type-out of a data region.
   Abort the current work. (Initialize after a set-up period).

4. In the event of an unrecoverable system error, the monitor provides for the saving of all data in the user's program and the closing of the data files. This prevents the user from losing raw data.

5. The monitor contains the device handler for the computer's real time clock.

6. The monitor allocates core memory and processing time in a multi-user system.

7. The monitor provides for its own initialization and the loading of the system loader.
8. The monitor provides system information to the rest of the system; e.g., user starting address, the first free location, the last free location, etc.

B. Handler

The experimental device handler, as a software module, plugs into the monitor. Because this software module is directly related to the characteristics of the particular experimental apparatus, it is important to understand its operation. The device handler attempts to accept the incoming data in a general yet efficient way, while leaving the user the maximum flexibility to manipulate raw data as required. A typical requirement is to take data from two ADC's and two live time clocks independently and simultaneously. The requirements here are in essence a subset of the more general I/O to a mass storage device. The experimental device handler, adheres to the following:

It is compatible with the monitor used. Additionally, the user is able to reassemble to be compatible with any monitor he may want to use.

If the device can alternatively use more than one interrupt facility, then the handler must be able to provide service either way.

Each interrupt that the device is capable of producing must have associated with it an appropriate service routine in the handler. The service includes both the treating of data and resetting the ADC's and clocks.

Specifically, the ADC's are treated as though they were a readonly mass storage device. The data is transferred from the ADC's directly to the PDP-15 memory via the 3-cycle data channel facility without program intervention. The handler is not accessed until N words (e.g., 32) have been transferred. The interrupt then means that 32 words have been transferred and the ADC's are now disabled. The service routine is accessed, the ADC's are enabled to fill a second 32 word list (which they automatically and immediately start doing), and the filled list is used to increment the appropriate channels in a histogram.

The description of a device handler falls into three categories; *data format*, *service routines*, and *input of information* from the program (user) to the handler.

1. The Data Format

The ADC words are transferred into a 32 word list in core. This list is a mix of data words from both ADC-A and ADC-B. The data for an event (single parameter) is represented by an 18 bit word. For a two parameter event, two words would be used. The 18 bit word is used as follows:

ADC data	13 bits
A or B	1 bit
Reject	1 bit
Reserve for routing or user tagging	3 bits

The live time clocks automatically count one in a fixed location in core for every clock tick.

2. The ADC Service Routine

The service routine must reset data acquisition to a second buffer so that more data can be acquired without program intervention.

The service routine empties the first list while the device fills the second list, but does not allow the device to interrupt again until the first list has been emptied (Figure 3);

To empty the list, it must successively fetch each word from the list and do the following:

Count the number of words fetched thus far and exit if this equals the size of the list; i.e., all words processed.
Test the word to determine if it came from ADC-A or ADC-B and branch to the appropriate routine.
Test for a reject and discard if it is a reject.
Substract a 13 bit window from the data word and discard if the result is negative.
Scale the resulting number to fit in the data region.
Mask out the unwanted bits.
Add the base address to get the channel address.
Count one in the channel and check for overflow.
On overflow, disable the overflowing ADC and inform the program by setting an overflow status flag.

The service routine may be assembled with sections optionally left out to eliminate some of the steps above that are not always necessary and that reduce speed such as substracting windows.

3. The Clock Service Routine

This service routine resets data taking. It does not allow the device to interrupt again until the service has been completed. The live time clocks should be set so that the service routine is accessed at an interval which corresponds to the minimum time the experimenter would like to resolve; e.g., 1 sec. The clock itself is typically much faster; e.g., 10 KHz.

The service routine additionally determines if the live time is being accumulated by the user. If so, then the appropriate clock register (an hour, minutes, seconds register) in the main program is incremented. If an overflow has occurred, the appropriate ADC is disabled and the clock overflow status flag is set.

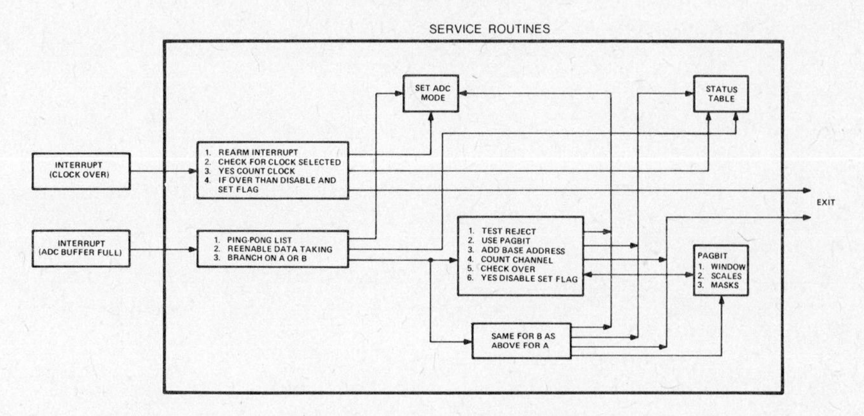

Fig. 3. A schematic of the experimental device handler's service routines.

4. Input Of Information From the User To The Handler

Communication between the user and the handler to input information to the handler is done indirectly via the monitor. The main program does not know the locations within the handler where the information is to be input. In the device independentenvironment the monitor maintains a table of this information and controls communication with the handler.

The user input of information to the handler for this example: (Figure 4).

a. Initialization: The device must be set-up for data acquisition which includes:

1. Set-up for the initial entrance to the service routine.
2. Preset the clock counter registers.
3. Set-up for the storage of overflow information.
4. Set-up of a system status table location.

b. Enter: A data table provided by the user is accessed and entered into the handler. In this case, this is a table of gain and window settings.

c. Close: The appropriate ADC is turned off.

d. Read: The appropriate ADC and clock is turned on and input is initiated. The base address and data mask are entered into the handler.

Throughout the experimental device handler each part is as modular as possible. A change in the form of the user input requires a corresponding change in the service routines. However, the user is free to add a subroutine to input additional data to the handler. The only changes that would be required would be in the service routines that get the new data. Changing the service routines can be easily

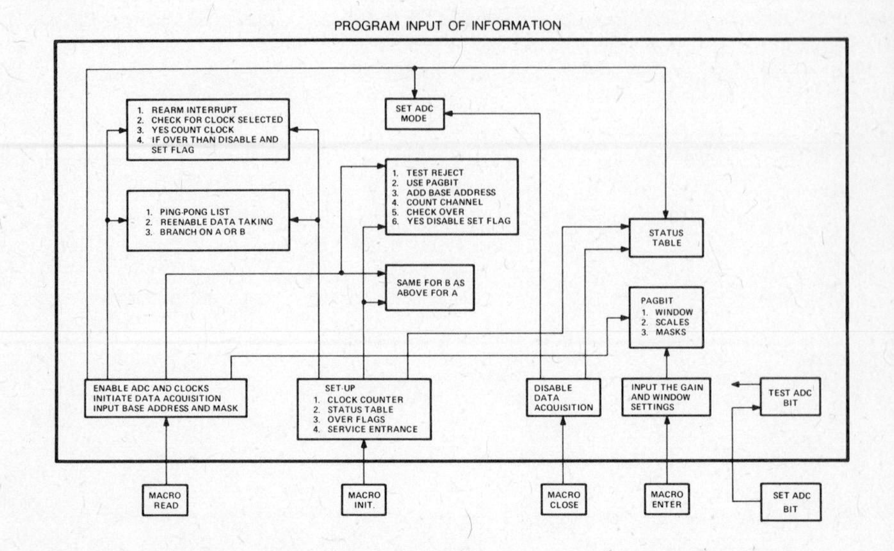

Fig. 4. A schematic of the user input of information to the experimental device handler.

accomplished since each section is as independent as is conceptually possible. That means, for example, that a branch on three ADC's instead of two would require a different branching routine and a new module to service the third branch. It is obvious that the addition of a third ADC would require some other changes, e.g., set mode and the status table would require alteration. A branch on the reject bit would not require any other changes in the handler except for the third branch routine.

C. Main Program

The main program plugs into the monitor and has the following functions:

It provides a framework for user command modules.

It provides a framework for user flag checking.

It controls data taking, data display, data manipulation, and data input/output.

A set of commands that are required for single parameter data acquisition is presented in Figure 5. Each of these commands represents a software module which plugs into the main program.

The command to be executed is input from the user. The device on which the input is made must be capable of alphanumeric input because of the potentially large number of commands and command arguments. The teletype is, therefore, a logical choice because all of the capability desired is already contained in the teletype handler which is a part of the monitor.

The main program must also contain an error diagnostic program to inform the user of what the nature of the error has been. The program checks overflow flags, dump flags, and the teletype flag, and outputs messages, displays, and allows for any additional user flags to be added. At the same time, the program remains modular in all aspects; any feature may be expanded and modified without restructuring the whole program.

COMMANDS

DATA TAKING

ENABLE
STOP
GAIN
WINDOW
ARM OVERFLOW
IGNORE OVERFLOW
CHARGE INTEGRATION
CHARGE OFF
PRINT CHARGE

CLOCK

LIVE TIME
REAL TIME
PRINT TIME

DATA MANIPULATION

HIGH POINT
INTEGRATION
BACKGROUND SUBTRACT
SUBTRACT DATA REGIONS
WRITE OUT CHANNELS
ZERO DATA

DISPLAY

DISPLAY
SET FULL SCALE
SET MARKERS
EXPAND VIEW
TAKE SQUARE ROOT DATA
DISPLAY LOG DATA
SQUARE ROOT OR LOG OFF
RESET
RESTORE
DISPLAY OFF
INTENSITY
CALIBRATE
DISPLAY TIME

FILE

OPEN
DUMP
ZERODUMP
AUTODUMP
ABORT
NEXT RECORD
READ NOTES
TYPE NOTES
CLOSE
DELETE
TEST PRESENT
RENAME

Fig. 5. Commands in a single parameter Pulse Height Analysis System.

The following approach is used. At the heart of the program is an executive loop. Whenever an operation in the loop has completed the program returns to the executive to go through the loop again. The loop involves the following (Figure 6):

1. Teletype Flag Interrogation

The first thing that is done in the loop is to check the teletype busy flag. If there is something to be done with the teletype, then the program proceeds to check the message output flag to determine if there is a message in the message output stack. If there is a message in the stack, then have the teletype handler output the message and return to the executive to start the loop again. If there is no message, then proceed to command decoding section.

2. Command Decoding

The command decoding section of the executive loop consists of four dispatch routines:

Dispatch #1. Here the commands that have been input by the user are fetched and each command is set up for the next dispatch routine. If all of the input commands have been processed, then an asterisk is output to the user to signify the readiness for his next set of commands. If an input command is found, the routine sets the loop to go to dispatch routine 2 on the next loop and then returns to the executive.

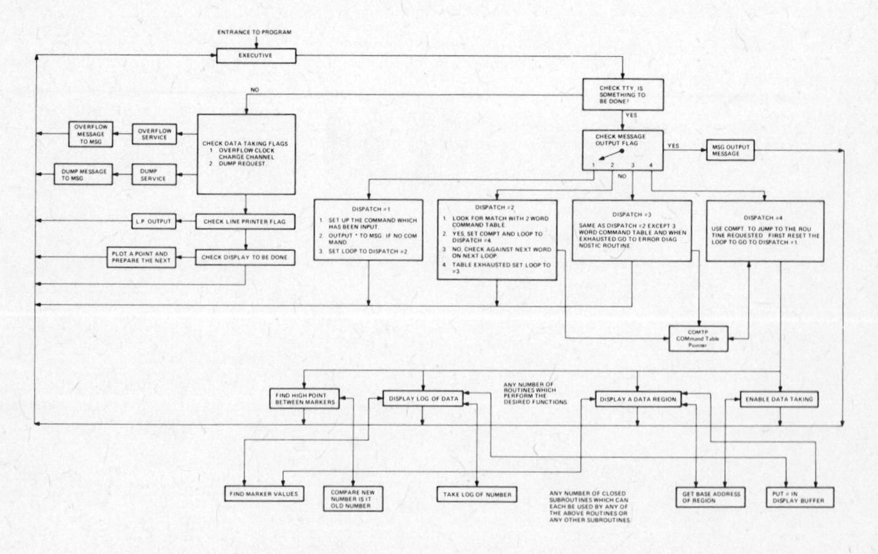

Fig. 6. A schematic of the user's main program.

Dispatch #2. Here the command is checked for a match with the two word command table and if a match is found, the entry point to the service routine is entered in the command table pointer register to be used by dispatch 4. On each pass through the loop, the command is checked against one entry and then the program returns to the executive. On finding a match, the loop is set to dispatch #4 for the next loop. On exhausting the command table, the loop is set to dispatch 3 on the next loop.

Dispatch #3. Here the command is checked for a match with the three word command table. The rest is the same as dispatch 2 except that if the table is exhausted then the entry for the error diagnostic routine is entered in the command table pointer register.

Dispatch #4. Here the loop is set to go to dispatch 1 on the next loop and then the program goes to the routine that will execute the input command.

3. Command Execution

The routines that perform the desired commands are open routines in that they are entered from dispatch routine 4 and always terminate with a return to the executive to go through the loop again. The routine itself can use as many closed subroutines as are required to execute the command. In general, anything that will be used by more than one routine is put into subroutine form and can then be used by as many parts of the program as the user desires. Entry to the subroutines is from anywhere in the program and return upon termination is always to the point from which it was entered.

Command input is inherently slow; there is no great requirement for speed in decoding. With this approach, it could take many passes through the loop before the command routine is accessed.

If at the start the teletype did not have anything to be done, then the program proceeds to check the data taking flags.

4. Data Flag Interrogation

The data taking flags are software status flags that inform the program that either an overflow has occurred or that a dump request has been made.

The overflow flags:

Clock overflow - The fact that an overflow has occurred is recorded in the notes files (mass storage, see Appendix) and a message is put into the output stack to inform the user that an overflow has occurred. Return is made to the executive.

Charge overflow - The ADC's are disabled (this is done via hardware for the clock overflow), the overflow is recorded in the notes file, and the user is informed via a message output on the teletype.

Channel overflow - The ADC is disabled by the handler, the overflow is recorded in the notes file and the user is informed that the overflow has occurred. If the user had previously chosen to ignore channel overflows, then the ADC is enabled and data taking is restarted, but the notes and message remain.

The dump requests: The user requests that data be automatically dumped on a given condition such as channel overflow. When this condition occurs, the data is recorded on mass storage in a data file with a particular record number. (See Appendix) The loop then proceeds to check any other flags that the user may wish to have checked. Here for instance the line printer output flag could be checked for an output message. The loop then goes to check for the display to be done flag. If a point is to be displayed, the program then displays it and prepares the next point for display. It then goes back to the executive to go through the loop again.

In the system under discussion, relative to the PDP-15 speed there is seldom something to do on the teletype, there is rarely an overflow or dump flag, and refreshing the display is the busy part of the program unless there is a storage scope which requires infrequent service. The executive loop, therefore, spends most of its time checking data taking flags, etc.

We have discussed the main program in detail. The point here is that the command modules plug into this program. It is important that commands may be changed very simply .

To add a new command (software module) the user adds the command and its pointer to the command decoder tables. He then plugs in the routine that executes the command.

The entrance is taken care of by dispatch 4.

The exit is taken care of by the standard return to the executive.

The routine can call for the use of any subroutine.

To add a new software flag, the user inserts the flag daisy chain style within the existing flags. He then plugs in the routine that services the flag.

The entrance is from the flag.

The exit is a standard return to the executive.

The routine can call for the use of any subroutine.

A multi-parameter program can be written as a straight forward extension of the single parameter program already described. The structure remains the same. The handler is altered. Many of the routines are modified, but these are easily added with a structure such as this.

D. Error Diagnostics

Error diagnostics are a necessary part of the main program structure. The user needs to know both that an error has been made in his command and also some details as to what he has done wrong. Examples of errors: No files open, data taking still enabled, illegal terminator, etc. When a new command is added, the corresponding diagnostic should be considered.

E. Assembly Time Options

Full utilization of the modularity requires that the user be able to chose among optional capabilities prior to assembly. This gives him the capability of having a library of routines that can be loaded into the program to suit his specific needs. Figure 7 shows some options of this nature. In addition to the options on the types of commands already discussed (that is data taking, clock, display, data manipulation, and file), the user has; hardware options that he will always use, (but he would like to be able to use someone else's routines created on a machine that doesn't have the option); hardware options that he will use rather randomly; and the option to run under different monitors or under the control of the dynamic debugging program. This gives the user an added dimension of modularity.

This then points to the next step. Which is that all of the routines that are options and in fact all routines that are not continually needed should reside on a disk. These routines would then be available to the user on request and with an all but negligible delay. The capability of his system would be greatly enhanced. This approach can be implemented now.

IV. SUMMARY

We have discussed modularity in data acquisition. Specifically, software modularity has been stressed.

The system discussed is centered around a monitor into which software parts can be interchangeably plugged. The monitor provides:

Operation in a relocatable environment.
Device Independence.
Accepts user input.

While operative, three main software modules plug into the monitor.

Experimental device handler.
User main program.
Peripheral device handlers.

The experimental device handler treats the user's data. Each section: (a) service routines, (b) user input of information; (c) data format; is as modular as possible, making user modification simple.

The main program plugs into the monitor and provides a nuclear physics framework into which user commands and user software flags can be plugged.

The final part is the software system with all of the programs required to create, debug and operate new user programs including all of the peripheral device handlers.

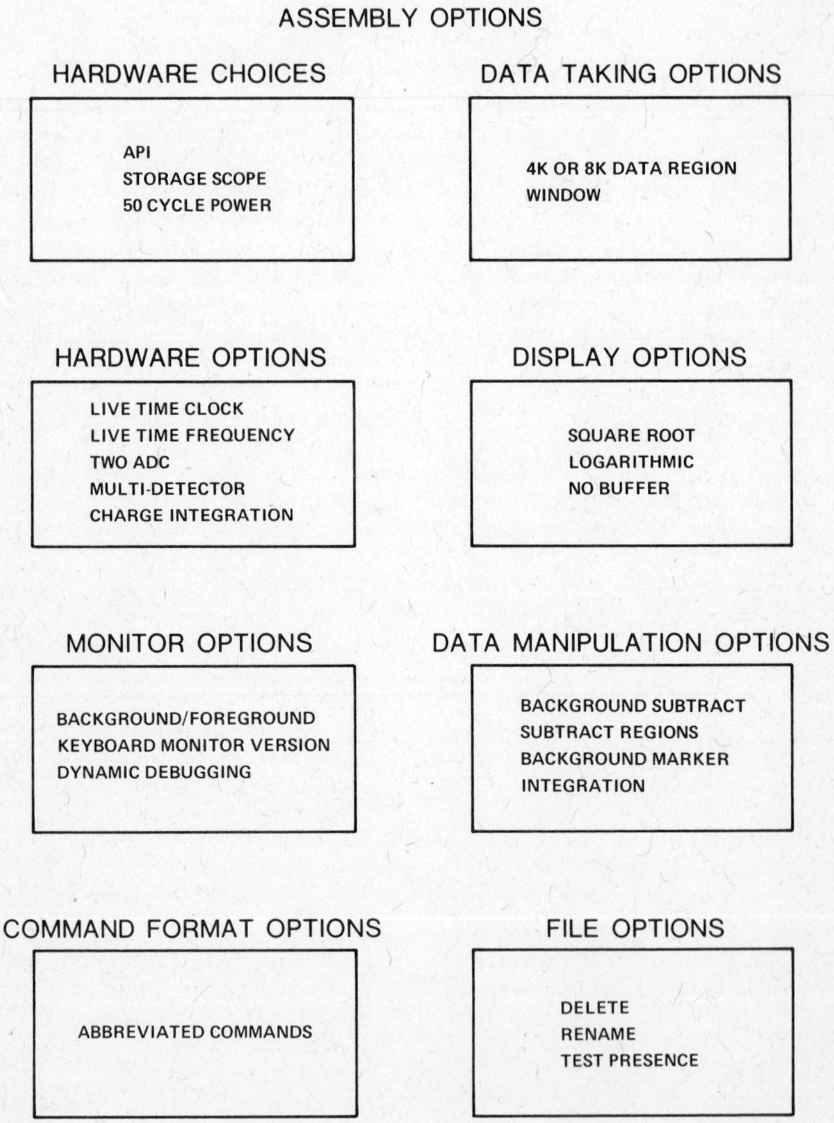

ASSEMBLY OPTIONS

HARDWARE CHOICES

API
STORAGE SCOPE
50 CYCLE POWER

DATA TAKING OPTIONS

4K OR 8K DATA REGION
WINDOW

HARDWARE OPTIONS

LIVE TIME CLOCK
LIVE TIME FREQUENCY
TWO ADC
MULTI-DETECTOR
CHARGE INTEGRATION

DISPLAY OPTIONS

SQUARE ROOT
LOGARITHMIC
NO BUFFER

MONITOR OPTIONS

BACKGROUND/FOREGROUND
KEYBOARD MONITOR VERSION
DYNAMIC DEBUGGING

DATA MANIPULATION OPTIONS

BACKGROUND SUBTRACT
SUBTRACT REGIONS
BACKGROUND MARKER
INTEGRATION

COMMAND FORMAT OPTIONS

ABBREVIATED COMMANDS

FILE OPTIONS

DELETE
RENAME
TEST PRESENCE

Fig. 7. Assembly options in a single parameter Pulse Height Analysis System.

Conclusion

With this approach, the user achieves the software modularity which allows him to quickly make the alterations required to go from one experiment to the next. The result is a reduction in the time and effort involved in such a change.

APPENDIX

Device Independence

Enough has not been said about device independence. The user should have the capability of assigning his output to any mass storage device. The appropriate device handler should then be loaded and available for his use. For any device, there should be the capability to substitute from a variety of device handlers which give a variety of capability. All the handlers should be standard and data or notes written with them should be accessible by the rest of the system. What sort of options will the user require? Different data formats must be accommodated (e.g., binary and ASC11), file oriented and non-file oriented data should be storable, a different number of files may be required at different times, writing or reading may be the only requirement. If there is one handler and it does all things for all people it will also be very uneconomical in its use of core. There should, therefore, be several handlers.
The user input of information to a mass storage device handler is usually more extensive than the input to the experimental device handler, discussed earlier. The following information is input:

INIT	The device and device handler are initiated.
READ	The data from the device is read into the user's program.
WRITE	The data from the user's program is written on the device.
WAIT	The user waits until the device is available.
WAITR	The user tests the availability of the device and either waits or goes away to return later.
CLOSE	This terminates an INIT SEEK or ENTER.
TIMER	This waits a specified time interval.
EXIT	A standard return to the monitor.
SEEK	Search the directory for a desired file.
ENTER	Used to make an entry in the directory.
FSTAT	Test for the presence of a file.
RENAME	Remove a file.
DELETE	Deletes a file.
TRAN	Reads or recores a block in a user specified format.

CLEAR Initializes file structure.

MTAPE Used for unique non-file oriented functions.

An example of the capabilities required for a particular device, DECtape, and the resulting device handlers are:

1. DTA (which requires 2290_{10} locations)

DTA is the most general DECtape handler available with the monitor. DTA has a simultaneous 3-file capacity, either input or output. Files may be referenced on the same or different DECtape units, except that two or more output files may not be on the same unit. All data modes are handled as well as all functions except MTAPE. Three 256_{10}-word data buffers are included in the body of the handler.

2. DTB (which requires 1548_{10} locations)

DTB has a simultaneous 2-file capacity, one input and one output. Both files may be on the same or different units. DTB transfers data only in ASCII or binary data modes. Included in the handler is space for two 256_{10} word data buffers. Functions included are:

INIT	ENTER	READ	WAIT	WAITR
SEEK	CLOSE	WRITE		

3. DTC (which requires 680_{10} locations)

DTC is the most limited and conservative in terms of core allocation) DECtape handler. DTC is READ ONLY handler with a 1-file capacity and only one 256_{10} word data buffer, it handles either ASCII or binary input. Functions included are:

INIT	CLOSE	WAIT	WAITR
SEEK	READ		

4. DTD (which requires 1560_{10} locations)

DTD has full function capabilities including MTAPE. It allows for only one file reference, either input or output at any given time. All data modes are acceptable to DTD. One 256_{10} word data buffer is included.

5. DTF (which requires 605_{10} locations)

DTF is a non-file-oriented, multi-unit DECtape handler which will accommodate, serially, up to eight DECtape units. This handler can perform both input and output in either ASCII or binary. DTF contains no internal buffers; the functions included in DTF are:

INIT	WRITE	MTAPE	WAITR
READ	CLOSE	WAIT	

User Commands In A Device Independence Environment

We have discussed device independence. Whereas this applies to all I/O, a more detailed discussion of the user commands for I/O of data to a file seems to be in order. These commands were presented in Table 1.
The storage of data on a mass storage device should be accomplished with simple commands. The data would be stored in a binary file and the experimental notes in an ASCII file. The notes consist of a record of all times, shut downs, charge accumulation and anything else that is of a nature that the experimenter wants to have recorded automatically plus anything that he wants to manually enter (type). All information, both data and notes should be recorded under a simple name of the user's choice and should subsequently be available to the user under the same name. The data and notes should be recorded by a standard device handler in a standard format. That is, they should be available to the rest of the system, e.g., an analysis program written in Fortran IV. The data file should contain as many records of data as the user requires. These should be accessible with the file name and record number.

Open Run 1	If Run 1 had been previously recorded, then a non-data taking file should be opened (i.e., the user cannot take data), the data read into the data region and the user informed via TTY that record #1 is in core. If Run 1 had not been previously recorded, then a data taking file should be opened. On the appropriate mass storage device, a binary file is opened for the data and on a second unit an ASCII file is opened for the experimental notes.
Dump	The data is dumped (completely independently of data acquisition) on mass storage and is accessible as a record number.
Zerodump	With the data taking disabled, the data is dumped on mass storage as a record and then the data region is zeroed.
Autodump	This sets up so that in the event of a channel overflow, the data taking is disabled, the data is dumped on a mass storage as a record, the data region is zeroed, and data taking is again enabled.
Abort	This is used to clear the data area, status flags, counters, etc. In essence, a fresh start discarding set up information and data .
Next Record	Used to examine data, this reads the next record or any record into core.
Read Notes	Used to examine the contents of the notes file when a non data taking file has been opened.

Type Notes	Used to insert user notes, non automatically, in the notes file when a data taking file is open.
Delete	Deletes both the binary data file and the ASCII notes file from mass storage.
Test	Tests for the presence of a file on mass storage.
Rename	Changes the name of the file from the old name to the new name.
Close	This closes both the binary and the ASCII files and then transfers the binary data onto the

same unit that the notes are on. The data taking file is no longer open, the user cannot take data or access the data he has just taken until he opens another file.

DISCUSSION

SEAMAN: Would you comment on how much the modularity concept adds to the memory requirements? Is the overhead on this like 10 or 20% more memory or what sort of figures are involved in that case?

PENCZER: Depends on what you are doing. If you consider that you would want all the routines that could be loaded in resonant, in core, you would require more memory than you do with the modularity approach which should be the case. If you want all of the options for outputing data to mass storage, it would be impossible without a modular approach. Certainly, if you load the options off a disc you reduce the core requirement considerably. To be specific, this program here would run on the 16K PDP 15. To run all of the options simultaneously, I really don't have a good feel, or you could certainly do less -- you can take data, you can output it to a mass storage device, you can display it in an 8K system, there are a lot of things you can do. This system here you can run, you can change the size of your data region with a non-modular approach. That is very difficult. I don't know if we answered the question or side-stepped it.

RUMMER (Chairman): I was wondering on the description of how the system operated, it would appear that you are asking the teletype is there anything to do and if there is, it gets taken care of, and then you go looking at the question of checking flags, does that suggest the teletype have priority over these things or did I misinterpret the picture at that point?

PENCZER: The teletype has priority in our setup over the other things, but the other things are not really data taking. The data taking is taken care of in the first modular device handler. Any data taking flag are responses in our channel overflow where you have the option in the device handler of also responding, such as turning data taking off. You have two ways of doing it: either always do it one way, which is in device handling, or we have to go back and program a change, or you can stop taking data for a second, roughly, depending on how many commands you have to go through in your dispatch routines, and then service an overflow. The only thing you have lost is that you haven't taken data for a second while you are waiting for the overflow. You have suspended data taking for that time.

RUMMER: What is a live time clock? Is that another word for real time, or does that have some other name?

PENCZER: That's an external clock, as far as we are concerned. It is configure from the ADC's to be running when the ADC's are enabled, and not running when they are not able. As far as we are concerned, it really does not make much difference how you have the clock set up. The real differentiation is between the computer's clock and the user's clock.

V.B. ADVANCING COMPUTER TECHNOLOGY AND IMPLICATIONS FOR NUCLEAR PHYSICS EXPERIMENTATION

by

J. Birnbaum
IBM Thomas J. Watson Research Center, Yorktown Heights, N. Y.

and

M. W. Sachs and D. A. Bromley
Yale University, Wright Nuclear Structure Laboratory
New Haven, Connecticut
(Presented by J. Birnbaum)

I. INTRODUCTION

The role of the digital computer in the low energy nuclear physics laboratory is now so established that at least three major conferences have been held solely on this topic.[1] This paper therefore is neither a review of an already well documented field nor a description of any specific system. Rather, it is an attempt to examine the possible effects of the rapidly advancing computer technology upon the manner in which experiments are conducted and, perhaps, the nature of the experiments themselves.

Any writer who attempts to predict the future must tread carefully between superficial platitudes and extrapolations unfairly biased by personal past experience. We will undoubtedly be guilty of both, but beg the reader's indulgence at the outset; our purpose is to impart some information about the directions in which computer systems appear to be growing and to express some thoughts as to the potential impact upon nuclear physics laboratories during the next decade.

II. SOME TYPICAL CURRENT USES OF COMPUTERS IN NUCLEAR PHYSICS

Examination of current usage of an on-line computer, drawn from three years experience at the Yale University Emperor tandem accelerator, will be useful as a reference point. The system, developed jointly by Yale and IBM Research,[2] has now been used in a great many experiments, most of them with the computer acting as an extremely sophisticated multi-parameter analyzer capable of analyzing data in real time. In some cases, closed- or open-loop on-line control of experiments has been used; data logging in all cases has been automatic. The system, as emphasized in earlier descriptions of its objectives and characteristics, has been designed to encourage highly interactive usage, and is powerful and flexible enough to be configured rapidly and operated in a natural and simple manner by physicists who are not computer experts. In these examples, applications in which the on-line computer makes a unique contribution are stressed, rather than those involving simpler (but still worthwhile) pulse height analysis. "On-line" here means both on-line to the experiments and on-line to the experimenter (in the sense of interactive programming).

1. Automatic Control of Experimental Conditions

Measurements of the $B^{11}(p,\gamma)C^{12}$ reaction have been made by Brassard et al[3] as part of a study of the structure of the C^{12} giant resonance. In these measurements, the gamma ray spectra are measured with large (5" x 6" and 9" x 12") NaI(Tl) counters. Since in these experiments only one in every 10^{10} photons from the target is of any real interest, and since such counters are subject to gain shifts which would significantly degrade the quality of the data or, indeed, render the experiment impossible, it is necessary to automatically monitor the gain and make the necessary corrections. The experimental setup is shown in Figure 1. Using the computer as a gain stabilizing system both allows a gain correction algorithm which is more elegant than that obtainable with conventional hardware stabilizers and also enables simultaneous monitoring of pileup and beam current adjustment to keep pileup within readily established and adjustable limits. The pulse height reference for stabilization is provided by a light pulser which is optically coupled to the photomultiplier. Prior to the start of the measurement, a light pulser spectrum is stored in the computer. During the measurement, light pulses are again accumulated and are routed to a spectrum separate from the data by the experimental counting equipment. Periodically, the new light pulser spectrum is compared to the original spectrum. By means of a least-squares fitting program, gain shifts <u>and</u> peak broadening are detected and measured. If a gain shift occurs, a correction is applied by transmitting the correction voltage digitally to a digital-to-analog converter connected to an experiment control and monitoring system,[4] which applies the necessary correction voltage to the control element of a variable high voltage supply in series to the main high voltage supply to the photomultiplier. If pileup is detected, through peak broadening, the accelerator operator is requested to reduce the beam current to an acceptable level. (The pileup referred to is the residual pileup remaining after processing by the pileup rejection circuitry in the fast electronics.) The loop can also be closed directly via DAC output to the accelerator ion source controls. This is an example of both open- and closed-loop feedback in one experiment.

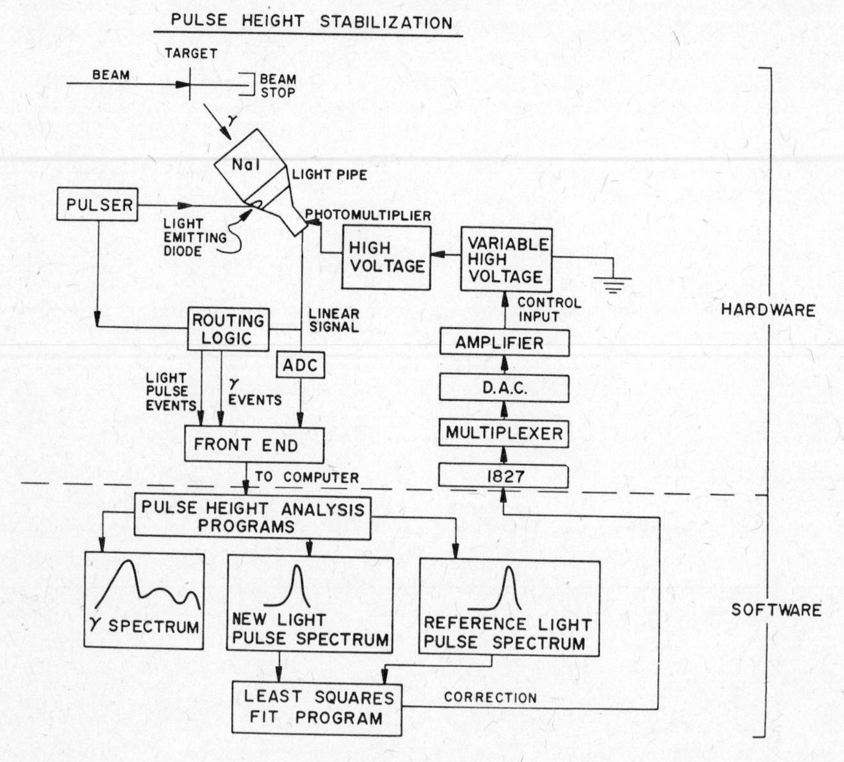

Fig. 1. Pulse height stabilization using the computer.

2. On-line Compensation for Physical Conditions

One of the experimental problems in particle-gamma angular correlation measurements in which the particles are detected at angles near 180 degrees is that the kinematic broadening of the peaks in the particle detector, reflecting the non-negligible angular width of the detector, can be enough to prevent resolution of the peaks of interest, especially when the reactions are induced with heavy ion projectiles. Use of the computer for data acquisition enables compensation for this effect by a method first used by Broude et al.[5] at the Chalk River Laboratories in Canada. In this method, particle detection is accomplished with an annular counter which is radially position-sensitive. The counter, as is usual in these experiments, is on the beam axis, the beam passing through the center hole. Both the energy and radial position (i.e., angle) of each particle are measured. Using the angle information, the computer corrects, event by event, the energies of the particles to correspond to a single angle of detection, thus removing the kinematic spread. Doing the correction event by event in real time enables the experimenter to see his spectrum with its corrected resolution as he accumulates it. Figure 2, taken from Ref. 5, illustrates the resultant dramatic improvement.

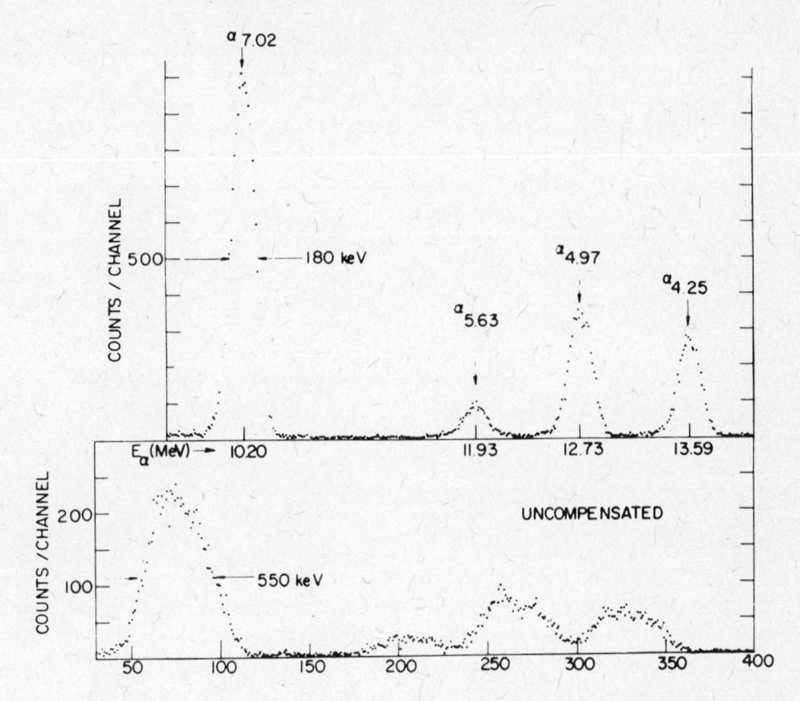

Fig. 2. Spectra of α-particles in time-coincidence with γ-rays are shown (lower) without and (upper) with on-line compensation for the angular dependent energy spread. From Ref. 5.

At Yale, the same approach has been extended through addition of a totally depleted transmission detector in front of the radially position-sensitive unit. This permits simultaneous identification of the particle involved through standard E x ΔE techniques and kinematic correction of the energy spectrum of each selected species. At the same time, coincident gamma radiation is detected in both LiGe and NaI detectors at appropriate angles surrounding the target and pulse height and time information from all detectors are recorded and processed for each event as it occurs. Currently, this system is being used in studies of reactions such as $Be^{9}(O^{16}, \alpha\gamma)Ne^{21}$; $Re^{9}(O^{16}, p\gamma)Na^{23}$, and $C^{12}(O^{16}, \alpha\gamma)Mg^{24}$ with concentration on the study of very high spin states and their electromagnetic de-excitation. Without an on-line computer, these studies would be difficult or impossible because, if the zero degree detector angular aperture were sufficiently restricted to retain adequate energy resolution to resolve groups feeding individual residual states, the corresponding solid angle would be so small that the particle gamma coincidence efficiency would be reduced below usable limits.

3. Interactive Data Reduction

In many present-day experiments, the parameter spaces encompassed by the measurements may be of the order of 10^6 or even 10^9 cells. For example, a typical particle-gamma coincidence experiment may involve recording of particle-gamma coincidences with particle resolution of 1024 channels, gamma resolution of 4096 channels, and fast coincidence timing with a resolution of 1024 channels, for a total of 4×10^9 cells. Even if memories of this size should become available, it is difficult to conceive of the human mind being able to comprehend this enormous volume of data when presented all at once. The typical measurement procedure, therefore, consists of recording each event on magnetic tape as it occurs, with its full resolution, while monitoring a selected subset of this parameter space in the form of one- or two-dimensional spectra. The first stage of data analysis then consists of replaying the event tape, performing pulse height analysis in one or two parameters while setting restriction gates on the remaining parameters. For example, a 4096 channel gamma ray spectrum may be accumulated in coincidence with a single particle group (i.e., range of particle pulse heights) and a particular range of fast coincidence times, perhaps with random coincidences (identified by their time information) subtracted. This process may be repeated many times for different particle groups.

This sort of analysis has been done for years using large off-line computers and programs using conventional sorting techniques. Experience at Yale shows that the technique becomes considerably simpler and more powerful when performed interactively with a program which treats the tape as if it were the original data source and allows the user to program as if he were doing real time data acquisition, or, in fact, allows him to use the same real-time program with which he originally accumulated the data. The display and other real-time facilities[6] enable him to make the frequently needed ad-hoc corrections to the analyses with the same ease with which the experimental apparatus is manipulated while accumulating data.

A case in point is the recent measurements of the reaction $C^{12}(O^{16}, \alpha_1)Mg^{24}(\alpha_2, Ne^{20})$ which were performed by Gobbi et al. at Yale.[7] One of the goals of this experiment is the measurement of spins and parities of levels in Mg^{24} by means of measurements of the angular correlations of the alphas from the decay of Mg^{24} to Ne^{20}. In the experiment, one alpha particle was detected at zero degrees by a particle identification telescope, and the other one was detected in a position-sensitive detector covering the range 20° to 90°. The experimental setup is shown schematically in Figure 3. Information recorded on magnetic tape for each

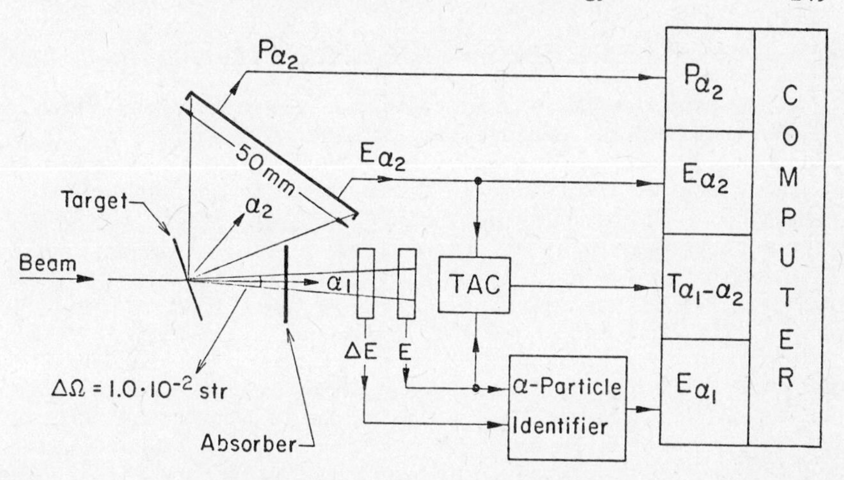

Fig. 3. Experimental setup for measurements of the reaction $C^{12}(O^{16}, \alpha_1)Mg^{24}(\alpha_2, Ne^{20})$.

event consisted of the energies of the two alpha particles, the position pulse height of α_2, and the timing information for the fast coincidence between the two detectors, each with 1024 channels resolution. The ultimate goal of the data reduction is to extract, for each level of interest in Mg^{24} (given by the energy of α_1), the angular distribution of α_2, going to the ground state of Ne^{20}. In the process of extracting the angular distributions, a number of corrections had to be applied. These were done in a very natural way with the aid of the display system.

It was first necessary to correct the energy of α_1 for a slow gain shift. Correction factors were obtained by extracting spectra of α_1 covering time intervals short compared to the drift times. Using these factors, the data were corrected, event by event, to a common reference gain. In making such a correction, it is necessary to account for the finite channel width in the data since the gain correction changes the widths also. The method used consisted of adding to the original channel numbers a random number between 0 and 1, multiplying by the gain correction factor, and then truncating the result to an integer. This is equivalent to temporarily transforming the data to a scale on which the channel widths are truly negligible. In the present case, it was adequate to assume a rectangular shape for the channel profile, but it would have been straightforward to fold in a shape factor with the random numbers. The output of this step was a corrected event tape which was used for all subsequent operations, illustrated in Figure 4.

The large variation of rise time of the pulses from the position-sensitive detector, as a function of position, precluded the simple setting of a digital gate on the time

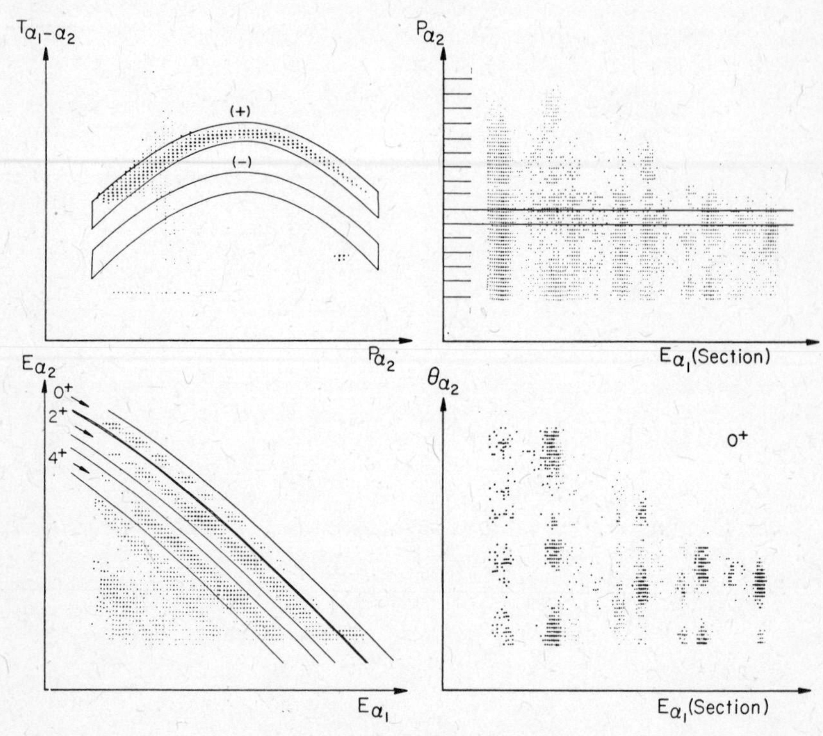

Fig. 4. Steps in reduction of the data on the reaction C^{12} $(O^{16}, \alpha_1)Mg^{24}(\alpha_2, Ne^{20})$.

information. It was therefore necessary to first replay the data to produce a two-dimensional spectrum of time as a function of the position pulse height from the α_2 (position-sensitive) detector (upper left in Figure 4). Using the light pen, a line was then drawn around the locus corresponding to the true coincidences. The experimenter then indicated the location of the random coincidences, and the computer marked an identical-shaped region there. The coordinates of the points on these lines were stored in the computer memory, and in subsequent steps each event was compared with the sets of coordinates marking the real and random coincidence regions. If an event fell inside the real coincidence region, a count was added in the final pulse height analysis; if it fell in the random region, a count was subtracted. After the time gates were set, a sample of the data was again replayed, subject to the time restrictions, and a two-dimensional spectrum of α_2 position as a function of α_1 energy was accumulated (upper right of Figure 4). The vertical loci shown represent angular distributions of the decay alphas from the various states in Mg^{24}. However, in this coordinate system, the alphas leading to the several low-lying states in Ne^{20} coincide, and it is necessary to perform another analysis to separate them.

Using the P_{α_2} vs. E_{α_1} plot, 14 ranges of P_{α_2} were selected and for each of these ranges, yet another replay was made, with all previous restrictions (discussed above) applied, producing the two-dimensional spectrum in the lower left of Figure 4, a plot of E_{α_2} v. E_{α_1}. In this coordinate system, the events corresponding to alphas going to the ground, 2^+ and 4^+ states of Ne^{20} are very well resolved, and the light pen may be used to indicate to the computer the region corresponding to, for example the ground state. Note that this region is a function of the position of α_2, and so it must be selected separately for each range of position.

Finally, the data were replayed one more time, applying simultaneously the two-dimensional gate region on time and P_{α_2}, and 14 gate regions (for different P_{α_2} ranges) on E_{α_1} vs. E_{α_2}. In addition, an event-by-event transformation of position into true angle was made, based on calibration data which were taken with a mask containing a set of calibrated slits over the detector. This resulted in a two-dimensional spectrum of α_2 angle as a function of E_{α_1} only for events in which the α_2 leads to the ground state of Ne^{20}. Finally, the appropriate vertical slices in this spectrum were summed to yield separated angular distributions of the ground state α_2 for each α_1 group (level in Mg^{24}).

4. Interactive Theoretical Calculation

Certain types of theoretical calculations, especially those involving parameter searches, are executed with vastly improved efficiency when the physicist's judgment and intuition can be invoked during the course of the calculation. For example, analysis of the data accumulated in measurements of the elastic scattering of protons by C^{12}, by M. J. LeVine and P. Parker[8] required a large number of phase shift fits to the angular distributions. Since a good set of initial values for the parameter search was not available, considerable exploration of the parameter space was required. Use of a conventional search program at the Yale Computer Center would have been prohibitively expensive, and funds notwithstanding, it is reasonably clear that even with unlimited time such a search would never have converged without direct physicist-computer communication during its course. Instead, the search was programmed for the 360/44 at the Wright Laboratory, with the results of each fit displayed on the cathode ray tube, and the physicist able to enter starting values for the next search through the function keyboard. In this way, it was possible to guide the program into the proper area of the parameter space in a much more efficient manner than would have been possible had the computer been working blindly along a pre-programmed trajectory. This direct coupling of the human intuition with the computational speed of a computer has been one of the long-term goals of the Yale/IBM project and one still under active study, as will be discussed in a later section.

5. Graphical Representation

Even in experimental situations which are amenable to treatment by older means, the computer can often provide considerable assistance simply in surveying the data, in ways which may foster new insights. For example, in the study of the elastic scattering of O^{16} by O^{16}, by J. Maher et al,[9] detailed excitation functions and a large number of angular distributions were taken at closely spaced energies. One of the mysteries of this type of work has always been the fact that the optical model parameters for single angular distributions may differ widely even at closely spaced energies. With the aid of a good automated graphics facility on the computer, it becomes possible to plot all the fits on common coordinates such that the major trends are immediately visible. In this case a pseudo three-dimensional plot of cross section as a function of energy and angle was made of all the data. The plot was made on the cathode ray tube and the plotting parameters were varied to obtain the best plot, after which it was drawn on the digital plotter for further study. An example of such a plot[10] is shown in Figure 5. Instead of a simple series of mountain ranges we see a number of saddle points reflecting transitions from one dominant partial wave in the scattering to that two units greater with corresponding transition from a $|P_L|^2$ to a $|P_{L+2}|^2$ angular distribution. While there is, in a sense, nothing new here, it may be claimed that the magnitudes and importance of these features were never realized before seeing the fits in this particular perspective.

III. THE NEW TECHNOLOGY

Computers have grown faster, smaller and cheaper at such an incredible rate during the last six years that it is interesting to project these advances into the next decade. Without question, new orders of magnitude in computing power will soon be available. The full impact of large scale integrated circuitry is still to be felt, and the effects upon the cost of computers, the number of instructions which can be executed per second and memory capacity will be enormous. The remainder of this paper attempts to examine some of the possible implications of these advances as they pertain to systems such as that described in the preceding section. Interestingly enough, developments in programming techniques and overall system organization may well be as important as the electronics improvements themselves.

Nuclear data acquisition systems, with few exceptions, have fallen into one of two categories. The first, and most common, is comprised of small, inexpensive and rather limited systems which replace hard-wired multi-parameter analyzers

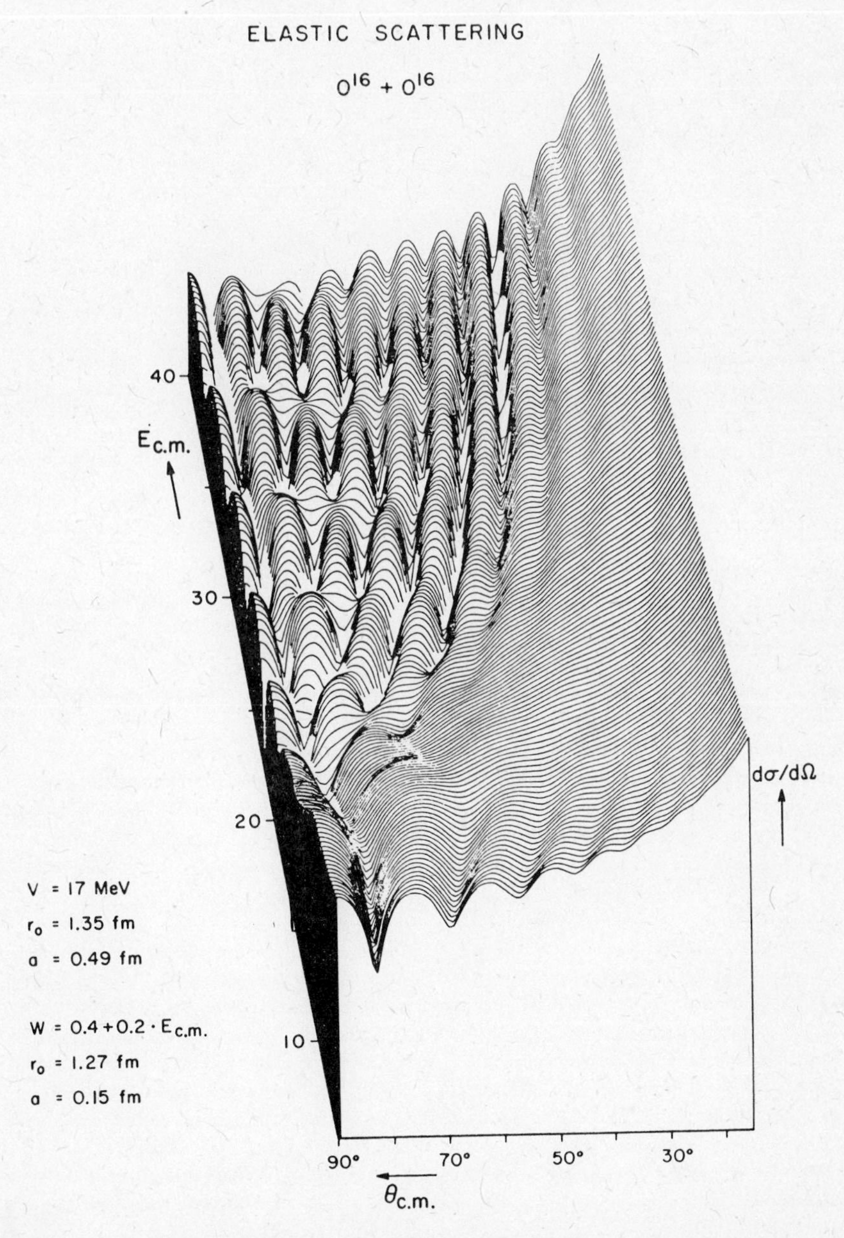

Fig. 5. Cross section of the elastic scattering of O^{16} by O^{16} as a function of angle and energy. Optical Model fits.

with the additional capability of providing simple control and limited data analysis. In the second category are those larger systems involving one or more intermediate sized computers, usually controlled by rather complex operating systems. These systems have data acquisition front ends which are either specialized hard-wired control units or mini-computers connected to a data channel or memory port of the computer. In both categories, many of the systems have worked well within their stated objectives. The quantity of data obtained, its quality, and the ability to store and access these data at a later time have been greatly enhanced. In almost every case, however, the systems have taken great effort to develop. Many are neither simple to program, nor simple to use once programmed. An interesting question to ask at this time then is whether the faster, smaller and cheaper computer systems now evolving will improve this situation. Are groups of these small machines likely to replace those machines now installed, or will laboratories move toward single, very powerful time-shared facilities? The most intriguing, and also most difficult, question is whether entirely new ways of doing nuclear physics, or, at least, of gathering, analyzing, and presenting the data, can be expected. A number of aspects of these questions are addressed below.

1. Memory

Our experience at the nuclear laboratories using the Yale/IBM system and at the IBM Research Center in a different laboratory environment, has been that the input-output capability of this system (a representative of the second of the two above mentioned categories) has been adequate to cope with all data rates and response requirements. However, with monotonous regularity, the memory capacity (32K, 32-bit words) of the computer has been exhausted. This is sometimes because of the need for very large (and perforce elaborately segmented) programs, and sometimes because large quantities of data must be buffered or analyzed.

It is certain that the emergence of very large, fast, inexpensive solid state memories will have great impact on all computer systems. Programs will be much easier to write, because segmentation in many cases will no longer be required, and because the operating systems under which the programs run should become much simpler. The principal function of any operating system is resource management, and many of the functions which are associated with present-day multiprogramming and time-sharing systems will become unnecessary as machines get faster and larger. Memory hierarchies, paging systems for time sharing, and the other devices invented by systems programmers to maximize resource utilization, should become less necessary. This can only be considered good news from the physicist's point of view.

More directly, applications programming will benefit. For example, the clever but elaborate techniques for associative analyzers which now, alone, permit mega-channel analyses should become unnecessary in most instances. Similarly, procedures such as automatic overflow tables for those cells of non-associative analyzers which have had more than a preset number of counts will also become superfluous. In short, the availability of large, inexpensive memories will lead to systems which are simultaneously simpler to use and more efficient.

2. Computing Power

The effects of the great increases in computing power are less straightforward to predict, but at the same time carry exciting promise. It is now commonplace to design machines capable of executing millions of instructions per second, and billion instruction per second processors are envisioned by some. This enormous computing power could be most effective in those experiments which benefit from extensive model calculations being performed in real time. For example, the work on reaction mechanisms through angular distribution studies, such as that of LeVine noted above, typically begins with an optical model fit. This often involves coupled elastic and inelastic scattering channels, and between 10 and 20 parameters is not unusual. One of the goals of such analyses is that of establishing the sensitivity of the model predictions to each parameter; with the current system, this is a tedious procedure, since several minutes of processor time are often required for each change. With the computer power soon to become available, it may become possible for the predictions to track the light pen-controlled parameter variations in real time. Exploration of a multidimensional surface would then become a far more tractable problem than it is now, and one which could provide increased insight into the physics involved.

There is, of course, great potential danger in this much power, as must be evident after a few moments reflection. The ability to do millions of numerical computations per second can never be a substitute for the analytical judgment of a trained scientist. One can picture unborn generations of nuclear physicists following the siren call into the swamps of numerical calculation. However, properly used this speed could have the effect of making the computer an extension of a physicist's best thinking with little or no effort on his part. More powerful systems could have the capacity to become relatively self-teaching. Conceptually, at least, far more difficult experiments could be attempted. For example, consider the search for super-transuranic elements, the current Holy Grail of nuclear experimentalists. Even the most optimistic estimates of production mechanisms in heavy ion fission reactions predict that the atoms of the super-transuranics, if produced at all, will be present in at best one in 10^{15} to 10^{16} of the heavy beam particles and other reaction products emergent from the target. The filtration and identification of these particles from the background represents a problem with many orders of magnitude more signal-to-noise ratio than has been attempted previously. No single current experimental technique can provide the necessary filtration, so that at the outset a series-parallel net of simultaneous measurements must be made on all accessible observables which could identify the new species (e.g., momentum, energy, time-of-flight, x-ray yield, fission yield characteristics, etc.). To achieve these simultaneously on a sufficient spectrum of the emergent particles will require a totally new generation of computational power and capacity.

3. Microprogramming

The influence of microprogramming--a technique whereby hardware instructions are themselves programmed in a set of more primitive operations--is becoming much more prevalent in modern computers. Microprogramming has great potential ramifications for nuclear physicists. In most contemporary machines, the instruction set is specified by the manufacturer but the advent of so-called writable control stores has made it possible in many cases for the user himself to alter the contents of the microprogram. Obsolescence in the face of new procedures, new peripheral devices, or changing interfaces is thus essentially eliminated. New ideas can be incorporated easily and machine instructions to perform unique functions can be created. A microprogram of the read-only variety offers far more secure protection than can a standard program executing from main storage, frequently an important attribute (e.g., accelerator control).

While it is glamorous to think of defining an exotic instruction set by means of microprogramming, it is, in fact, unlikely that very many such applications will see fruition. The real power of the technique lies in the ability to define controllers adaptive to the specific and peculiar needs of I/O devices. Thus, for example, high efficiency interfaces to another computer, a disk drive, an external memory, or a multiparameter analyzer could all use an identical input/output processor (channel) but with different resident microprograms. In fact, this is already being done in a few commercial products. At the moment, microprograms are difficult to write but as these techniques become more common higher-level, user accessible microcode generators will probably become available.

Microprogramming will also affect the organization of systems. The most common current approach for the multiexperiment (or, in come cases, multi-task) environment is that of minicomputers surrounding a centrally shared intermediate or large computer. This approach, as has been pointed out elsewhere,[11] has the concomitant disadvantages of higher cost and lower performance when compared with specialized control units connected directly to data channels. Microcoded controllers can be expected to replace, in many instances, the small general-purpose machines now being used as specialized front ends. Performance should increase and costs decrease. A prevalent system organization may be a group of these input/output processors connected to a central machine. The individuality, flexibility, and preprocessing capability of the mini-to-central processor organization could then be combined with the cost and performance advantages of the hard-wired control unit approach.

4. System Organization

Systems in which all input/output devices are connected in a loop such that data transmission is unidirectional but at such speeds that all response requirements can be satisfied are now available for factory automation. Other systems are available or have been proposed which permit direct connection to a bi-directional data buss so that all devices connected to the buss are modular, thereby allowing information to be transferred directly from device to device, without involving channels or the computer memory. This should see enormous application in all data acquisition applications because data can often be logged directly from the experiment onto tape or disk. Similarly, one can expect to see such utility operations as tape-to-printer and disk-to-plotter being performed with minimal overhead through direct connection of those devices (perhaps under the supervision of a microprogrammed input/output processor).

5. Communications and Networks

Most current remote access to computers is over low-speed voice grade telephone lines. There are, of course, many exceptions, but, in general, expensive equipment and complex programming packages for error checking and recovery are required. The communications field is rapidly changing, and one hears much talk of nation-wide networks of heterogeneous computers. Ideally, the user of such a facility will present his requirement to a network supervisor computer which after an examination of the available resources will select the computer or computers best able to handle the job, schedule the work, receive the output, and route it back to the appropriate terminal, all without intervention or knowledge of the user. A system representing an early version of such facilities is now being developed at the IBM Research

Center. The Yale Nuclear Structure Laboratory will be one of the nodes in this network. Yale will be connected over telephone lines to a System/360 Model 50 which serves as the system supervisor. Initially, the line speed and system responsiveness may limit Yale's use of the network to large-scale backup computing; conceptually however, there is no reason why other computers could not be called in during the course of an experiment and this is clearly one of the goals of the continuing joint study. If this can be accomplished, it will have great importance, particularly for small laboratories.

6. Interfacing: Hardware and Software

Probably the most successful aspect of the Yale/IBM collaboration has been the ease with which the system has been used by non-expert personnel. The interface between the transducers and digitizers and the computer is a data highway with a standard set of signal characteristics. This concept of a modular set of digital instrumentation has now been extended and generalized by the ESONE committee of Euratom, as the so-called CAMAC standard,[12] which is now gaining wide acceptance. It seems likely that, in time, computer manufacturers will support the interface with programming systems and, in some instances, interface controllers built into the computers.

Another interface which has made the Yale system easy for physicists to use is that between the experimenter and the computer. Advanced display techniques, incorporating extensive use of light pens and function keyboards, have facilitated the selection of experimental alternatives during an experiment and have eliminated most of the training usually required for an operator of such a system.[6] The extensive use of higher level languages, particularly a superset of Fortran, and underlying data structures which enable run-time alteration of experimental parameters have proven well worth the effort expended in their creation.

There remains, however, a great deal which could be done in the area of user-oriented languages. Interactive higher level program preparation, either on dedicated machines, or in partitions of time-shared or multiprogrammed systems would be very useful. Systems today use either pre- or post-compilers, as at Yale, or extensive macro-assembly facilities. Such procedures are very wasteful of processor time and memory, and an extensible language which could be efficiently specialized to a given application would be a great improvement. This is a hard job, but many people are working on it, and it is quite possible that we shall see a data acquisition and process control language standard evolve. An obvious benefit, besides those already mentioned, would be the enhanced sharing of real-time programs among laboratories with dissimilar processors.

One of the more unique features of the IBM/Yale Joint Study is that it has developed into an informal collaboration among three nuclear physics laboratories and IBM. The other two are the NASA Space Radiation Effects Laboratory (a synchrocyclotron) at Newport News, Virginia, and the University of Maryland Cyclotron Laboratory (a azimuthally-varying-field cyclotron). Each has an on-line IBM 360/44 computer system with the same front end and display terminal as are used at Yale. All are using the same software and are jointly developing the next generation of this software. Because of the extreme generality of both the front end and the display system, this standardization is possible in spite of the quite different experimental conditions prevailing at the three quite different accelerators. In addition, S.R.E.L. functions as a service facility for a numver of user laboratories. Members of these laboratories perform experiments at S.R.E.L. and take their data home for analysis. In many cases, the data are in the form of event tapes (raw data) and are analyzed on the users' own computers, using the same software system (but on different System/360 models) as was used to record the data. It is quite possible that widespread use of the CAMAC standard together with source language computer independence will make this type of standardization feasible on small systems as well as the relatively large machines involved in the Yale/IBM project.

IV. A PROTOTYPICAL SYSTEM

An experimental system embodying some of the concepts and procedures enumerated in the preceding sections, but using contemporary hardware and software technology has been constructed at the IBM Watson Research Center. The system will be dedicated to full-time support of diverse experiments in January 1971, and will be reported on in detail at the forthcoming Nuclear Science Symposium.[13] The system design has not been optimized for the nuclear environment; nevertheless, one of its users will be a 3 MeV Van de Graaff facility which will be time-shared with solid state laboratories, analytical chemistry instruments, lasers, biophysics experiments, and high-speed synchronous devices such as scanning electron microscopes.

The system represents an attempt to exploit the commonalities of real time systems both at the computer and user interfaces. Thus, list oriented data structures are provided for the dynamic interaction with application programs. Higher level language facilities and facilities for the creation of these languages are provided. The system achieves efficiency by performing what are normally high overhead software tasks (such as interrupt response, and task scheduling) with special purpose hardware. "Intelligence" is distributed among the data channel, a special control unit, a communication system which propagates a complex interface to laboratories as far as one mile from the computer site, and a small computer-like local controller which drives a generalized CAMAC interface to the experiments themselves.

In essence, this system attempts to provide an organization which appears to the user as a dedicated front end computer which coordinates the gathering of data and the distribution of control signals to the experimental apparatus with a network of backup computers. An effectively infinite resource of computing power is available, but with increasingly poorer response characteristics as one moves farther from the experiment. The heart of the Yorktown system is again a 360 Model 44 computer, but with the data channels significantly augmented. The incremental cost of connecting new experiments to the system is small since the expensive facilities are shared among all users.

As in the Yale system, the overriding principle has been to keep the data channels running continuously with full responsibility for the input and output of information, with central processor interference being kept to a minimum. Functions such as polling of the laboratories, responding to interrupts, scheduling of tasks, simple error correction, and basic pre-processing (such as limit checking, event counting, etc.) are relegated to the channels and associated special purpose hardware. Overhead will be lessened still further by provision of device-to-device data transfer capability, should that prove necessary. The system has been constructed of currently available circuitry and with common programming techniques, but it is our expectation that experience gained during its operation will provide useful input to designers of future data acquisition systems.

V. CONCLUSIONS

Although on-line utilization of digital computers in nuclear physics is a relatively new phenomenon, it has already had far-reaching consequences for the field. The emerging computer technologies, hardware and software, seem certain to extend this influence. Contrary to widely held misconceptions, the major impact will be that of greatly improved quality of experimental data, and of increased scope of nuclear experimentation rather than of simply increased quantity of data, although the latter is also important. The heightened data quality will permit more stringent testing of hypotheses than now possible, and may make feasible the search for extremely elusive nuclear events. The Promise of large-scale, shared, remote facilities may significantly extend the useful research lifetimes of smaller and older research facilities. Paradoxically, these new systems should be proportionately less expensive and easier to use than their predecessors.

REFERENCES

1. *Proceedings of the Conference on the Utilization of Multiparameter Analyzers in Nuclear Physics*, L. Lidofsky, Ed., Columbia Univ., CU(PNPL)-227 (1962); *Proceedings of the Conference on Automatic Acquisition and Reduction of Nuclear Data*, K. H. Beckurts et al., Eds., Gesellschaft für Kernforschung m.b.H. Karlsruhe, Germany (1964); *Proceedings for the Skytop Conference on Computer Systems in Experimental Nuclear Physics*, Columbia Univ., EANDC (U.S.) 121 U, (1969).

2. H. L. Gelernter et al., Nucl. Inst. and Meth. 54, 77 (1967); J. Birnbaum and H. Gelernter, IEEE Trans. Nucl. Sci., NS-15, 147 (1969); J. Birnbaum and M. W. Sachs, Physics Today, July 1968, p. 43; J. Birnbaum, *Proceedings for the Skytop Conference on Computer Systems in Experimental Nuclear Physics*, Columbia Univ., EANDC(U.S.) 121U, 91 (1969).

3. C. Brassard, doctoral dissertation, Yale Univ. 1970; C. Brassard, J. P. Coffin, W. Scholz, H. Shay, and D. A. Bromley, Nucl. Phys., to be published.

4. C. E. L. Gingell, M. W. Sachs, D. A. Bromley, A. A. Guido, and J. Birnbaum, IEEE Trans. Nucl. Sci. NS-16, 165 (1969).

5. C. Broude and R. W. Ollerhead, Nucl. Inst. Meth. 41, 135 (1966).

6. J. Birnbaum et al., IBM J. of Res. & Dev. 13, 52 (1969).

7. A. Gobbi, P. R. Maurenzig, R. G. Stokstad, and R. Wieland, submitted to the Houston Meeting of the American Physical Society, Oct. 15-17, 1970.

8. M. J. LeVine and P. Parker, Phys. Rev. 186, 1020 (1969).

9. J. V. Maher, M. W. Sachs, R. H. Siemssen, A. Weidinger, and D. A. Bromley, Phys. Rev. 188, 1665 (1969).

10. A. Gobbi, private communication.

11. J. Birnbaum, *Proceedings for the Skytop Conference on Computer Systems in Experimental Nuclear Physics*, Columbia Univ. EANDC(U.S.) 121U, 91 (1969).

12. ESONE Report EUR 4100e, March 1969.

13. J. Birnbaum, Invited paper, IEEE Nuclear Science Symposium, New York, Nov. 4-6, 1970 (to be published in IEEE Trans. Nucl. Sci.).

DISCUSSION

SANTO: I want to ask you about alpha correlation experiments. Did you make a division of the position signal of the position sensitive detector within the computer?

BIRNBAUM: Yes. You mean the 14 division? That whole operation is done in replay from the correct events taped.

SANTO: And this was fast enough because you will also have the elastically scattered heavy ions coming into this counter?

BIRNBAUM: I don't believe that I am familiar with that work. The experiment will be discussed at the Houston meeting. I think Peter Parker will give the talk. [Bull. Am. Phys. Soc. 15, 1654 (1970).]

SEAMAN: You talked about distributing intelligence out to these various on-line aspects -- how different is that, from say to the model 44 interface at Yale?

BIRNBAUM: Very different. Completely different, I would say. The concept is the same, the implementation is very different in the sense that we take a channel, the data channel - IBM's very limited version of an input-output processer -- and build special hardware onto it which augments that interface in the sense that it becomes completely programmable and can do many things which previously it couldn't. For example, we can write channel/sub-programs which branch and link to themselves. We can do conditional transfers in the channel. We can link in real time between the CPU and the channel with that linkage initiated from either side. Instead of the six channel commands which are in the 360 repertory, as you buy a system 360, we have 226, and they involve mostly control operations but also a great deal of special controls which gets passed through the first level and out into the local controllers which are in the laboratory so that for example, if we want to do something like stopping or starting a bank of scalars or ADC's, or what have you, this is done not from the central processor but rather from the channel, but as interpreted at whatever the appropriate level of control down the line. The reason you think about doing this now is that hardware is really so terribly cheap. We are talking about a few cents a circuit that you can think of things like polling laboratories, by the way -- that happens in a nanosecond time scale in this particular system, that it is quite reasonable to think of eliminating the schedule components of the operation system in favor of some sort of hardware device which will do this sort of stuff. The other thing about that is the problem with such a system and the reason why I have been reluctant to talk about it publicly until now is that it would seem we are creating a system which instead of being easier to use is becoming a system which is more complicated, more and more computers to worry about and more and more interfaces. And so what we have done is to create what we call a special interface assembler which treats the input-output system as if it really were a second processer, and then presents to the user a very simple high level super FORTRAN set of statements in which he programs and then that gets translated by two different levels of language processors in the channel programs. The difficulty is that the channel programs are being changed in real time as the load on the system changes, so you have to be able to modify channel programs on the fly and yet guarantee people that you won't be disturbing their experiments, so the monitors are quite complicated but they are all working now. We don't have much feedback on how easy it is to use but it is certainly easier to use than the Yale system. Quite a bit easier, as a matter of fact.

SEGEL: Our big problem with coincidence experiments is the replay time. That's really what is bad because it's just data sitting around, you just don't ever get around to looking at it because you don't want to take the immense amount of time to replay the tapes over and over again. Is there anything on the horizon that will make life easier?

BIRNBAUM: Yes, I think that one of the things that perhaps we can look for are displays which have local processing capability in them, so that if you could do this device-to-device you could think of taking the event tapes and for some cases, at least doing much of the buffering and processing locally to a remote display, which hopefully would be a video or something of that sort which would be inexpensive enough that you wouldn't have to tie up the computer with replay. Yale has just installed, and I don't know if its operational yet, one of these ARDs terminals, which is the Tektronik's storage scope, key board, and function keys, and they are going to be attempting to do a lot of non-interactive data analysis and program operation using that remote terminal becuase the computer scheduling time at Yale has become 24-hr/day, 7-day week propositions with theorists relegated to after midnight and Saturdays and Sundays, and things of that sort.

RUMMER (Chairman): Do you view the process control language as an extension of FORTRAN?

BIRNBAUM: I feel very non-confident about that. We are using FORTRAN because most of our people are familiar with it and it's available. We have written a family of processors. We have a compiler which compiles pre-compilers which then go into the FORTRAN compiler, so it is very simple for us to create a special language based on Fortran, and since we only have a few programmers it's just a matter of means.

We do have a large group working on what we call extensible languages and we talk to them. They are very abstract at the moment. There is a language which was just called to my attention called PROTRAN which was developed in England, which I just saw a citation on, which is an attempt at a universal process control data acquisition language with a FORTRAN-like structure and has global symbols and other facilities that a lot of people have talked about for a long time, but I think it is going to be awfully hard and a very speculative item, to talk about a language standardization.

LIST OF PARTICIPANTS

R. G. Arns	The Ohio State University
H. Bakhru	State University of New York
G. B. Beard	Wayne State University
R. C. Bearse	University of Kansas
J. A. Becker	Lockheed Missles and Space Co.
J. Birnbaum	IBM, Inc.
P. Brussaard	Duke University
R. T. Carpenter	The University of Iowa
B. Castel	Queen's University
C. C. Chang	Stanford University
C. M. Class	Rice University
J. P. Davidson	University of Kansas
J. Dubois	Chalmers University of Technology
F. E. Dunnam	University of Florida
F. E. Durham	Tulane University
P. M. Endt	Rijksunversiteit, Utrecht
G. A. P. Engelbertink	Brookhaven National Laboratory
S. Fiarman	University of Kansas
R. W. Finlay	Ohio University
H. T. Fortune	University of Pennsylvania
P. Goldhammer	University of Kansas
E. Halbert	Oak Ridge National Laboratory
S. S. Hanna	Stanford University
G. I. Harris	Aerospace Research Laboratories
M. Harvey	Chalk River Nuclear Laboratories
H. Hennecke	Aerospace Research Laboratories
R. Horoshko	University of Rochester
A. J. Howard	Trinity College
A. K. Hyder	Aerospace Research Laboratories
B. D. Kern	University of Kentucky
N. Koller	Rutgers University
R. W. Krone	University of Kansas
H. M. Kuan	University of Kansas
T. T. S. Kuo	State University of New York
D. Kurath	Argonne National Laboratory
R. Lawson	Argonne National Laboratory
J. R. MacDonald	Bell Telephone Laboratories
S. Maripuu	Rijksunversiteit, Utrecht
C. E. Moss	University of Colorado
S. A. Moszkowski	University of California
R. Penczer	Digital Equipment Corp.
F. W. Prosser, Jr.	University of Kansas
A. Rollefson	University of Notre Dame
M. L. Roush	University of Maryland
D. I. Rummer	University of Kansas
R. Santo	Universitat Munchen
L. W. Seagondollar	North Carolina State University
G. G. Seaman	Kansas State University
R. E. Segel	Argonne National Laboratory

R. Socash	Digital Equipment Corp.
D. R. Tilley	North Carolina State University
E. Titterton	Australian National University
J. Umbarger	University of Kansas
E. K. Warburton	Brookhaven National Laboratory
B. H. Wildenthal	Michigan State University
L. Zamick	Rutgers University